I0446530

Structural Biology of Membrane Proteins

A Text-Based Informative Book

Rickbed Nandi

Preface

In the empire of modern biology, the intricacies of life are woven into a complex tapestry of molecules, mechanisms, and interactions. Among these, membrane proteins stand as sentinels at the heart of this intricate web, governing a wide array of vital processes within the cells of all living organisms. The structural biology of membrane proteins, an ever-evolving and dynamic field, embarks on a journey to unravel the enigmatic world of these proteins, shedding light on their secrets and, in turn, transforming our understanding of life itself.

This book is a testament to the relentless pursuit of knowledge, the spirit of discovery, and the tireless dedication of countless scientists who have embarked on this journey. Its chapters are designed to serve as a comprehensive guide, offering readers a structured, insightful, and engaging expedition through the fascinating terrain of membrane protein research.

Our journey begins with an introduction, providing the reader with a foundational understanding of membrane proteins, their significance, and the challenges inherent in their study. From classification to isolation and expression, subsequent chapters delve into the fundamental techniques and methods employed in the pursuit of membrane protein structural elucidation.

The core of the book navigates through the heart of structural biology, exploring the tools and technologies that enable researchers to probe the atomic structures of these proteins. From X-ray crystallography to NMR spectroscopy and cryo-electron microscopy, each chapter unravels the power and limitations of these techniques, illustrated with compelling case studies and breakthroughs.

Beyond the confines of structural determination, we venture into the dynamic world of membrane protein function. These proteins are not static entities but dynamic players in cellular processes. We discuss their roles in health and disease, their interactions with lipids, and the intriguing potential for drug targeting and therapeutic design.

Our expedition continues with a closer look at the intricate processes of membrane protein folding, assembly, and engineering. Through these chapters, we peek into the emerging world of membrane protein manipulation and the prospects for creating designer proteins with tailored functions.

Ethical considerations are an essential aspect of our journey. We reflect on the ethical imperative that accompanies scientific discovery, emphasizing the responsible conduct of research and the equitable sharing of knowledge.

Throughout our voyage, we emphasize the interdisciplinary nature of membrane protein research. The crossroads of bioinformatics, computational modelling, and the integration with systems biology all serve as guiding stars in our quest to unlock the secrets of these essential proteins.

The promise of the field unfolds as we discuss the emerging technologies that are shaping the future of membrane protein research. Nanodiscs, lipidic cubic phase, advanced labelling techniques, and the vast potential of lipidomics usher us into an era of unprecedented possibilities.

In the final chapters, we step back to take a panoramic view of the field's potential, not only in the scientific domain but also in the broader context of society. Membrane protein research has the potential to redefine biomedicine, inspire innovation, and facilitate transformative breakthroughs.

As the journey concludes with our reflections on the limitless potential of this field, we invite readers to join us in the contemplation of the uncharted territories and the beckoning frontiers that lie ahead. The structural biology of membrane proteins is not just a scientific endeavor; it is an exploration of life itself. It is an ode to the curiosity that drives us, the passion that fuels our inquiry, and the ethical compass that steers our course.

The pages that follow are an invitation to embark on a voyage of discovery, an opportunity to grasp the secrets hidden within the cell membranes, and a window into a world where scientific inquiry meets the profound mysteries of life. The structural biology of membrane proteins is a remarkable tale of human ingenuity, boundless potential, and the ceaseless quest to unveil the secrets of the living world. It is our privilege to share this narrative with you, dear reader, as we embark on a captivating journey through the structural biology of membrane proteins.

Dedication

This book is dedicated to the students and general audience

Contents

Chapter 1: Introduction to Membrane Proteins

1.1 What are Membrane Proteins?

Membrane proteins are enigmatic molecular entities that play pivotal roles in the life processes of cells and organisms. Situated at the boundary between the interior and exterior of biological cells, these proteins serve as dynamic gatekeepers, selectively allowing the passage of essential molecules while guarding against the entry of harmful substances. Membrane proteins are the guardians of cellular life, orchestrating crucial functions such as signalling, transport, adhesion, and energy conversion. In this chapter, we embark on a journey to explore the multifaceted world of membrane proteins, delving into their classification, structural diversity, and functional significance.

The Crucial Role of Membrane Proteins

To grasp the fundamental significance of membrane proteins, one must first appreciate their omnipresence in the biological empire. These proteins are found in the membranes of all living cells, from the simplest bacteria to the most complex multicellular organisms. Their omnipresence underscores their indispensability in maintaining cellular integrity and enabling life-sustaining processes.

Membrane proteins serve as molecular gatekeepers, controlling the entry and exit of various molecules into and out of cells. This regulation is vital for the maintenance of cellular homeostasis, as it ensures that essential nutrients, ions, and signalling molecules are transported into the cell, while waste products and toxic substances are kept at bay. For instance, ion channels, a subclass of membrane proteins, facilitate the passage of ions (e.g., sodium, potassium, calcium) through the cell membrane, thereby maintaining the electrical and chemical gradients necessary for nerve signalling and muscle contraction.

In addition to their roles as gatekeepers, membrane proteins are integral to cellular communication. Receptors embedded in the plasma membrane, such as G-protein-coupled receptors (GPCRs), are responsible for detecting extracellular signals, including hormones, neurotransmitters, and light. When a specific molecule binds to these receptors, it triggers a cascade of intracellular events, ultimately leading to cellular responses. For instance, the binding of adrenaline to its receptor on the surface of heart muscle cells initiates a signalling pathway that increases heart rate and contractility, preparing the body for a "fight or flight" response.

Moreover, membrane proteins are involved in cell adhesion and recognition. Proteins like cadherins and integrins facilitate cell-cell adhesion and interactions with the extracellular matrix, playing pivotal roles in processes such as tissue development, immune response, and wound healing.

Structural Characteristics of Membrane Proteins

Membrane proteins are a structurally diverse group, but they share a common feature: their intimate association with the lipid bilayer that forms the cell membrane. This association is a result of the amphipathic nature of membrane proteins, with hydrophobic regions that embed within the hydrophobic core of the lipid bilayer, and hydrophilic regions that interact with the aqueous environment inside and outside the cell.

Two major categories encompass membrane proteins based on their relationship with the lipid bilayer: integral and peripheral membrane proteins.

Integral Membrane Proteins

Integral membrane proteins, also known as transmembrane proteins, traverse the lipid bilayer, extending from one side to the other. These proteins are firmly anchored within the membrane, and their hydrophobic transmembrane domains typically consist of α-helices or β-barrels. Integral membrane proteins can have single or multiple transmembrane segments, depending on their function and the number of hydrophobic regions required for insertion into the lipid bilayer.

A prime example of an integral membrane protein is the sodium-potassium pump (Na+/K+ pump), essential for maintaining the electrochemical gradients of sodium and potassium ions across the cell membrane. This transmembrane protein has multiple α-helical segments that span the lipid bilayer, allowing it to actively transport sodium ions out of the cell and potassium ions into the cell against their respective concentration gradients, a process critical for nerve impulse transmission and muscle contraction.

Peripheral Membrane Proteins

In contrast to integral membrane proteins, peripheral membrane proteins are not deeply embedded within the lipid bilayer. Instead, they are associated with the membrane's surface, typically through electrostatic interactions with the polar head groups of lipids or other membrane proteins. Peripheral membrane proteins can be released from the membrane without disrupting the lipid bilayer's integrity.

An illustrative example of a peripheral membrane protein is the protein kinase C (PKC), involved in cellular signalling pathways. PKC is initially cytoplasmic but becomes activated and translocates to the inner leaflet of the plasma membrane in response to specific signals. There, it interacts with phospholipids and other membrane-associated proteins to initiate a phosphorylation cascade, regulating various cellular processes.

Diversity in Membrane Protein Functions

The remarkable diversity of membrane proteins arises from their multifunctional roles in cell physiology. These proteins are tailored to perform specific functions essential for cellular survival and adaptation to changing environments. Here are a few examples that highlight the broad spectrum of their functions:

Transport Proteins: Membrane transporters, such as ion channels and transport carriers, facilitate the movement of ions, nutrients, and other molecules across the membrane. For instance, glucose transporters (GLUTs) are integral membrane proteins responsible for glucose uptake into cells, a fundamental process in energy metabolism.

Receptors: Receptor proteins on the cell surface recognize extracellular signals, enabling cells to respond to changes in their environment. For instance, the insulin receptor, a transmembrane protein, binds insulin, initiating a signalling cascade that regulates glucose uptake and metabolism.

Enzymes: Some membrane proteins are enzymatic and catalyse biochemical reactions at the membrane interface. Cytochrome P450 enzymes, found in the endoplasmic reticulum's membrane, play a crucial role in drug metabolism and detoxification.

Structural Proteins: Integral membrane proteins can contribute to the structural integrity of cells and tissues. For example, connexins, a family of transmembrane proteins, form gap junctions that allow direct communication between adjacent cells in tissues like the heart and brain.

Adhesion Proteins: Membrane proteins like cadherins and integrins mediate cell-cell and cell-extracellular matrix adhesion, critical for tissue development and immune response.

Therefore, membrane proteins are a diverse and functionally versatile group, integral to numerous biological processes. Their roles extend beyond mere structural components of the cell membrane, as they actively participate in maintaining cellular homeostasis, responding to external signals, and executing a wide array of biochemical reactions.

In this introductory section, we have embarked on a voyage of discovery into the world of membrane proteins. We have established their central role in governing the life processes of cells, highlighted their structural diversity, and showcased their multifunctional significance in various biological contexts. As we progress through this book, we will delve deeper into the intricate mechanisms that underlie the structure and function of membrane proteins, exploring the methodologies employed in their study, the challenges faced by researchers, and the profound impact of membrane protein research on fields ranging from medicine to biotechnology. Our journey into the structural biology of membrane proteins promises to unveil the mysteries of these vital cellular components and their pivotal contributions to life itself.

1.2 Importance of Membrane Proteins

Membrane proteins are remarkable molecular entities that play pivotal roles in a wide array of cellular processes, making them indispensable components of life itself. Their unique location within cellular membranes grants them a distinct set of functions, and as we delve into the importance of membrane proteins, we uncover their fascinating contribution to biology, medicine, and biotechnology. In this section, we explore the multifaceted significance of membrane proteins through both theoretical understanding and concrete examples.

Facilitators of Cellular Transport

At the heart of every living cell lies the essential process of cellular transport. Membrane proteins serve as the gatekeepers, orchestrating the movement of ions, molecules, and even larger substances across the hydrophobic barrier of the cell membrane. Take, for instance, ion channels. These specialized membrane proteins create selective pathways that allow specific ions to flow across the membrane, regulating the electrical potential of cells. Voltage-gated sodium channels, found in nerve cells, are prime examples. When these channels open, they permit the rapid influx of sodium ions, initiating the transmission of electrical signals—a process critical for nerve cell communication and muscle contraction.

In addition to ion channels, transporters represent another vital class of membrane proteins. These proteins facilitate the active or passive transport of a wide range of molecules, including nutrients, neurotransmitters, and drugs. The glucose transporter GLUT1, found in red blood cells, ensures a continuous supply of glucose, a primary energy source, to fuel cellular activities. Without such transporters, cells would be isolated and starved of essential nutrients.

Receptors for Cellular Signalling

Membrane proteins also function as receptors, relaying extracellular signals to the cell's interior, a process crucial for cellular response and coordination. G-protein-coupled receptors (GPCRs), a prominent class of membrane receptors, exemplify this role. These receptors span the membrane seven times and bind to ligands—hormones, neurotransmitters, or even photons of light. Upon ligand binding, GPCRs undergo conformational changes, activating intracellular signalling pathways. For example, the beta-adrenergic receptor, a GPCR, responds to adrenaline, initiating a cascade of events that increase heart rate and contractility—a response vital for the body's "fight or flight" mechanism.

Additionally, receptor tyrosine kinases (RTKs) are another class of membrane proteins with significant importance. These receptors, found on the surface of cells, bind growth factors and activate cellular growth and differentiation pathways. An illustrative case is the epidermal growth factor receptor (EGFR), which, when mutated or overexpressed, can contribute to the development of various cancers. The therapeutic targeting of EGFR with drugs like Erlotinib has revolutionized cancer treatment, emphasizing the profound significance of membrane proteins in medicine.

Structural Support and Cellular Adhesion

Beyond their roles in signalling and transport, membrane proteins contribute to the structural integrity of cells and tissues. Integral membrane proteins, such as cadherins, form adherens junctions, which are crucial for cell-cell adhesion and the maintenance of tissue architecture. Cadherins play an instrumental role in embryonic development, where precise cell adhesion and movement are required. For instance, E-cadherin is indispensable for the formation of the blastocyst during early embryogenesis in mammals.

Furthermore, membrane proteins are essential in anchoring cells to the extracellular matrix. Integrins, a family of transmembrane receptors, mediate cell-matrix adhesion, enabling cells to adhere, migrate, and interact with their surroundings. Disruptions in integrin function can lead to a myriad of pathological conditions, including autoimmune diseases and cancer metastasis.

Enzymatic Catalysts within Membrane

While enzymes are often associated with soluble proteins, membrane proteins also harbour enzymatic activity, and these enzymatic functions are pivotal for various cellular processes. Cytochrome P450 enzymes, embedded within the endoplasmic reticulum membrane, are responsible for metabolizing drugs, toxins, and endogenous compounds in the liver. Their role in drug metabolism has vast implications for personalized medicine, as genetic variations in these enzymes can influence an individual's drug response and susceptibility to adverse effects.

Additionally, ATP synthase, a remarkable membrane protein complex, plays a central role in cellular energy production. Located in the inner mitochondrial membrane, it couples the proton gradient generated during respiration to the synthesis of ATP, the cell's primary energy currency. This enzyme powers nearly all cellular processes, underlining the critical importance of membrane proteins in cellular bioenergetics.

Targets for Drug Discovery and Therapeutics

One of the most tangible and impactful aspects of membrane proteins' importance is their relevance in drug discovery and therapeutic interventions. A significant proportion of pharmaceuticals target membrane proteins, making them invaluable in the pharmaceutical industry. GPCRs, in particular, are an attractive target class for drug development due to their role in a wide range of diseases, from hypertension to neurological disorders. The development of β-blockers targeting GPCRs, which revolutionized the treatment of hypertension and heart disease, stands as a monumental achievement in medicine.

Similarly, ion channels represent another class of membrane proteins with therapeutic potential. Calcium channel blockers, which modulate the activity of calcium ion channels, are widely used to treat conditions like hypertension and cardiac arrhythmias. Voltage-gated sodium channels are also the target of drugs used in the management of pain and epilepsy.

Membrane proteins involved in infectious diseases offer another avenue for therapeutic intervention. The HIV protease, an essential membrane protein in the HIV life cycle, has been targeted by antiretroviral drugs like ritonavir, leading to significant advancements in the treatment of HIV/AIDS.

Membrane proteins are integral components of cellular life, orchestrating a wide array of functions that span from cellular transport and signalling to structural support and enzymatic catalysis. Their significance in biology, medicine, and biotechnology cannot be overstated. Membrane proteins are not only the architects of life at the molecular level but also the prime targets for innovative drug discovery and therapeutic interventions. As we journey deeper into the world of membrane protein structural biology, we will uncover the intricate mechanisms that underlie their functions and the profound impact they continue to have on our understanding of life and health.

1.3 Challenges in Studying Membrane Proteins

Membrane proteins, as integral components of cell membranes, play a pivotal role in various biological processes, including signal transduction, ion transport, and molecular recognition. Their essential functions make them prime targets for scientific inquiry and drug development. However, the study of membrane proteins is fraught with unique challenges that have long perplexed researchers. In this section, we delve into the formidable obstacles encountered when investigating these vital biological players, shedding light on the intricate nature of membrane proteins and the innovative strategies devised to overcome these hurdles.

Hydrophobic Environments and Solubilization

One of the foremost challenges in studying membrane proteins is their inherent hydrophobic nature, which renders them insoluble in water—a medium commonly used in traditional biochemical experiments. Membrane proteins are designed to reside within lipid bilayers, where their hydrophobic regions are shielded from the aqueous environment. Consequently, when extracted from their native membranes, they tend to aggregate or misfold due to exposure to water, impeding their structural and functional characterization.

Example: Rhodopsin, a membrane protein responsible for vision in humans, is notoriously challenging to solubilize. Its hydrophobic transmembrane domains make it prone to aggregation when removed from the lipid environment.

Researchers have had to develop innovative methods, such as the use of detergent micelles, to maintain its solubility for structural studies.

To address this challenge, researchers have developed a range of detergents specifically formulated to solubilize membrane proteins while preserving their structural integrity. Detergents create micellar structures, encapsulating the hydrophobic regions of the proteins, allowing them to remain stable in aqueous solutions. Popular detergents include n-dodecyl-β-D-maltoside (DDM) and lauryl maltose neopentyl glycol (LMNG). However, selecting the appropriate detergent is often a trial-and-error process, as different proteins may require distinct conditions for solubilization.

Membrane Protein Expression and Production

Producing sufficient quantities of membrane proteins for research purposes is another significant hurdle. Unlike their soluble counterparts, which can be expressed in bacterial systems like Escherichia coli, membrane proteins often demand more complex and time-consuming expression systems. This complexity arises from the need to mimic the native lipid environment, including the lipid bilayer, during protein production.

Example: The cystic fibrosis transmembrane conductance regulator (CFTR), a chloride ion channel, requires specialized cell lines or systems, such as mammalian cells or insect cells, for expression. These systems can closely replicate the native environment, ensuring proper folding and post-translational modifications necessary for functional studies.

Furthermore, some membrane proteins are toxic to host cells, further complicating expression. Researchers have developed strategies such as co-expression with chaperones or the use of lower expression temperatures to mitigate toxicity.

The optimization of membrane protein expression systems has also led to the development of fusion tags and modification of protein sequences to enhance stability and solubility during expression. These modifications often include the addition of affinity tags for purification or the use of fusion proteins to aid in folding.

Structural Heterogeneity and Conformational Flexibility

Membrane proteins are dynamic molecules, exhibiting conformational changes essential for their functions. This inherent flexibility poses a significant challenge to structural studies, as it can lead to the capture of multiple conformations or prevent the determination of a stable structure. Membrane proteins can adopt various functional states, and capturing these states at atomic resolution is a formidable task.

Example: The adenosine A2A receptor, a G protein-coupled receptor (GPCR), undergoes substantial conformational changes upon ligand binding. These changes are crucial for signal transduction but complicate structural studies. Researchers employ techniques like cryo-electron microscopy (cryo-EM) to capture distinct conformational states.

Innovative approaches, such as lipidic cubic phase crystallization and nanodisc technology, have been developed to stabilize membrane proteins in their native lipid environment, allowing researchers to explore their structural diversity. Additionally, advancements in cryo-EM have enabled the visualization of membrane proteins in various conformations, shedding light on their dynamic nature.

Lack of High-Resolution Structural Data

Compared to soluble proteins, the number of high-resolution structures available for membrane proteins is significantly limited. This dearth of structural data stems from the challenges associated with their study, including difficulties in crystallization and the determination of their structures at atomic resolution.

Example: While there are thousands of structures available for soluble proteins in the Protein Data Bank (PDB), membrane protein structures represent a small fraction. This imbalance underscores the need for continued efforts to develop novel techniques and technologies.

To address this limitation, researchers have turned to alternative structural biology methods, such as solution NMR spectroscopy and cryo-EM, which have become increasingly popular for membrane protein structural studies. These methods allow for the determination of structures in native-like lipid environments, circumventing some of the challenges posed by traditional crystallography.

Integration of Structural and Functional Data

Understanding the relationship between membrane protein structure and function remains a complex puzzle. While structural biology provides crucial insights into protein architecture, correlating this information with functional data presents challenges.

Example: The mechanosensitive ion channel MscS is an example of a protein where structural insights alone cannot fully elucidate its function. Combining electrophysiological recordings with structural studies has been essential to understanding how MscS responds to mechanical stress.

Researchers are increasingly adopting an integrative approach that combines structural techniques with functional assays, computational modelling, and bioinformatics analysis. This holistic approach allows for a deeper understanding of membrane protein function and aids in the identification of potential drug-binding sites and allosteric regulation.

Thus, the study of membrane proteins is a dynamic and evolving field that continues to grapple with numerous challenges. These challenges arise from the unique properties of membrane proteins, such as their hydrophobicity, dynamic nature, and complex expression requirements. Overcoming these obstacles requires the concerted efforts of researchers from diverse disciplines, as well as the development of innovative techniques and technologies. Despite the challenges, the pursuit of membrane protein structural biology remains essential for advancing our understanding of cellular processes and for the development of novel therapeutics targeting these vital proteins.

1.4 Historical Overview of Membrane Protein Structural Biology

In the complex needlepoint of biology, membrane proteins are the sentinels, the gatekeepers that allow for the precise orchestration of biochemical processes across cellular boundaries. Understanding the structure of these enigmatic molecules has been a pursuit of great significance, marked by remarkable milestones and pioneering breakthroughs. In this section, we embark on a historical journey through the annals of membrane protein structural biology, tracing its evolution from its nascent stages to the cutting-edge techniques of today.

The Dawn of Membrane Protein Structural Biology

The inception of membrane protein structural biology can be traced back to the early 20th century when the study of biological membranes and their constituents was in its infancy. At this stage, scientists primarily relied on biochemical methods and electron microscopy to explore the architecture of cellular membranes. It was a time when the molecular intricacies of these membranes remained concealed, awaiting the advent of technologies that could unravel their secrets.

The Rise of X-ray Crystallography

A watershed moment in the history of membrane protein structural biology came with the introduction of X-ray crystallography. This powerful technique, pioneered by Max von Laue and William Henry Bragg, allowed researchers to examine the atomic and molecular structure of crystalline substances. In 1952, Linus Pauling and Robert Corey proposed the concept of the alpha-helix, a groundbreaking insight that laid the foundation for understanding the secondary structure of proteins.

However, the study of membrane proteins remained elusive due to the challenges of crystallizing them. Membrane proteins are inherently hydrophobic, residing within lipid bilayers, and their extraction and purification posed significant obstacles. As a result, progress in this field was slow during the early days of X-ray crystallography.

The Myoglobin Breakthrough

A significant turning point came in 1958 when John Kendrew and his team successfully determined the three-dimensional structure of myoglobin, a water-soluble globular protein found in muscle tissue. This accomplishment, for which Kendrew was awarded the Nobel Prize in Chemistry, showcased the potential of X-ray crystallography in solving complex biological structures.

The myoglobin structure, with its helical segments and heme-binding site, demonstrated the power of structural biology in deciphering the architecture of proteins. It fuelled optimism that similar approaches could be applied to membrane proteins, igniting interest in their structural elucidation.

The Rhodopsin Revelation

The quest to crack the structural code of membrane proteins took a significant leap forward with the determination of the rhodopsin structure in 2000. Rhodopsin, a light-sensitive receptor found in the retina, plays a pivotal role in vision. The achievement was a testament to the perseverance of scientists and the continuous evolution of technology.

This breakthrough was led by Georgios Vassilakis and Nigel Unwin, who employed electron microscopy to unveil the structure of rhodopsin at a resolution of 4.5 angstroms. Their work revealed the seven-helix transmembrane structure of rhodopsin, providing a blueprint for understanding the architecture of G-protein coupled receptors (GPCRs), a superfamily of membrane proteins involved in numerous cellular processes.

The GPCR Renaissance

The elucidation of the rhodopsin structure catalysed a renaissance in GPCR research, sparking interest from pharmaceutical companies seeking to develop drugs targeting these proteins. In 2007, Brian Kobilka and Robert Lefkowitz received the Nobel Prize in Chemistry for their pioneering work on GPCRs, underscoring the importance of understanding these membrane proteins.

The GPCR structural revolution continued with the determination of the crystal structures of several GPCRs, including the beta-adrenergic receptor, the adenosine A2A receptor, and the dopamine D2 receptor. These structures not only deepened our understanding of GPCR function but also facilitated structure-based drug design, leading to the development of novel therapeutics for various diseases, including asthma and hypertension.

The Emergence of Membrane Mimetics

While X-ray crystallography and electron microscopy provided essential insights into membrane protein structure, the challenge of obtaining well-ordered crystals for all membrane proteins remained a bottleneck. Researchers began exploring alternative approaches, such as the use of membrane mimetics.

Detergents, liposomes, and nanodiscs emerged as valuable tools in membrane protein research. These substances provided a more native-like environment for membrane proteins, enabling their extraction, purification, and structural analysis. Notably, nanodisc technology, developed by Stephen G. Sligar, offered a novel means of stabilizing membrane proteins for structural studies.

The Cryo-Electron Microscopy Revolution

The recent history of membrane protein structural biology is marked by the cryo-electron microscopy (cryo-EM) revolution. This technique, which involves flash-freezing samples to preserve their native state, has overcome many of the limitations associated with traditional X-ray crystallography.

In 2013, the Nobel Prize in Chemistry was awarded to Richard Henderson, Jacques Dubochet, and Joachim Frank for their pioneering work in developing cryo-EM. Cryo-EM has since become a game-changer in the field, enabling the determination of high-resolution structures of challenging membrane proteins, including ion channels and transporters.

The Genomic Era and Beyond

The completion of the Human Genome Project in 2003 ushered in the genomic era, providing researchers with a wealth of genetic information. This vast resource, combined with advances in structural biology techniques, has accelerated the pace of membrane protein research.

Today, the field of membrane protein structural biology continues to evolve rapidly. Cryo-EM has reached unprecedented levels of resolution, enabling the study of dynamic conformational changes in membrane proteins. New computational methods and bioinformatics tools have emerged to assist in the prediction and analysis of membrane protein structures.

Hence, the historical journey of membrane protein structural biology is one of perseverance, innovation, and collaboration. From the early days of X-ray crystallography to the cryo-EM revolution, scientists have tirelessly strived to unlock the mysteries of these vital cellular components. As we move forward, the promise of new technologies and interdisciplinary approaches holds the potential to reveal even deeper insights into the structure and function of membrane proteins, further enriching our understanding of life at the molecular level.

Chapter 2: Membrane Protein Classification

2.1 Types of Membrane Proteins

Membrane proteins are a diverse group of biomolecules that play crucial roles in the structure and function of cell membranes. These proteins can be classified into several distinct types based on their structure, function, and location within the lipid bilayer. Understanding the different types of membrane proteins is essential for grasping the complexity of cellular processes and their relevance to various biological phenomena.

Integral Membrane Proteins

Integral membrane proteins, also known as transmembrane proteins, are a prominent class of membrane proteins that traverse the lipid bilayer, with portions of their structure embedded within the hydrophobic core of the membrane. These proteins are integral to the integrity and functionality of biological membranes and serve a wide array of functions.

Examples of Integral Membrane Proteins

G-Protein Coupled Receptors (GPCRs): GPCRs are integral membrane proteins that play a pivotal role in signal transduction. They respond to a variety of extracellular signals, including hormones and neurotransmitters, and activate intracellular signalling pathways. For instance, the beta-adrenergic receptor, a GPCR, regulates heart rate and blood pressure by responding to adrenaline.

Ion Channels: Ion channels are membrane proteins that create selective pathways for the passage of ions across the membrane. These proteins are crucial for maintaining the electrical potential and ionic balance of cells. The voltage-gated sodium channel is an example, involved in the initiation and propagation of nerve impulses.

Transporters: Membrane transporters are integral membrane proteins responsible for the active or passive movement of ions, nutrients, or other molecules across the membrane. The glucose transporter GLUT1 facilitates the transport of glucose across the cell membrane.

Receptor Tyrosine Kinases (RTKs): RTKs are transmembrane proteins that play a role in cell signalling and regulation of cellular processes like growth and differentiation. The epidermal growth factor receptor (EGFR) is an example of an RTK that regulates cell proliferation.

Peripheral Membrane Proteins

Peripheral membrane proteins are associated with the membrane but do not penetrate the lipid bilayer. They are often attached to the membrane surface through interactions with integral membrane proteins or lipids. These proteins perform various functions and are generally soluble in aqueous solutions.

Examples of Peripheral Membrane Proteins

Guanine Nucleotide Exchange Factors (GEFs): GEFs are involved in the regulation of small GTPases, which play a key role in cellular signalling. GEFs, such as SOS (Son of Sevenless), are known to bind to the plasma membrane, facilitating the activation of Ras GTPases.

Peripheral Enzymes: Some enzymes are anchored to the membrane by binding to specific lipids or integral membrane proteins. For example, phospholipase C (PLC) hydrolyses phosphatidylinositol 4,5-bisphosphate (PIP2) in the plasma membrane.

Cell Signalling Proteins: Proteins involved in cell signalling, such as protein kinase C (PKC), often have domains that bind to specific lipid molecules in the membrane. PKC is activated upon binding to diacylglycerol (DAG) and calcium ions, both present in the membrane.

Single-Pass and Multi-Pass Membrane Proteins

Membrane proteins can also be categorized based on the number of times they traverse the lipid bilayer. This classification gives rise to single-pass and multi-pass membrane proteins, each with distinct structural and functional characteristics.

Examples of Single-Pass and Multi-Pass Membrane Proteins

Single-Pass Membrane Proteins: These proteins traverse the lipid bilayer only once. An example is the insulin receptor, which spans the membrane once and contains an extracellular domain for insulin binding and an intracellular domain for signalling.

Multi-Pass Membrane Proteins: Multi-pass membrane proteins traverse the lipid bilayer multiple times. The neurotransmitter transporter, serotonin transporter (SERT), is a multi-pass membrane protein that facilitates the reuptake of serotonin from the synaptic cleft into neurons.

Glycoproteins in the Membrane

Many membrane proteins are modified with carbohydrate chains, making them glycoproteins. These glycans can be attached to the extracellular domains of membrane proteins and play crucial roles in cell recognition, adhesion, and signalling.

Examples of Glycoproteins in the Membrane

Glycosylated Receptors: Several cell surface receptors, such as the epidermal growth factor receptor (EGFR), are glycoproteins. Glycosylation can influence receptor stability and ligand binding.

Immunoglobulin Superfamily (IgSF) Proteins: Many IgSF proteins in the immune system are glycoproteins. For instance, CD4, a glycoprotein, is a co-receptor for the HIV virus.

Cell Adhesion Molecules: Proteins involved in cell-cell adhesion, like selectins and integrins, are often glycoproteins. They mediate interactions between cells and the extracellular matrix.

Membrane proteins represent a diverse class of biomolecules with various structures and functions. Integral membrane proteins, peripheral membrane proteins, single-pass, multi-pass proteins, and glycoproteins all contribute to the complexity and functionality of cellular membranes. Understanding these different types of membrane proteins is essential for unravelling the intricate processes that occur within and around cell membranes. In the subsequent chapters of this book, we will delve deeper into the techniques and methodologies used to study and characterize these proteins, providing a comprehensive overview of the structural biology of membrane proteins.

2.2 Integral Membrane Proteins vs. Peripheral Membrane Proteins

While signifying the excellence of membrane proteins, a fundamental classification scheme distinguishes between two primary categories: integral membrane proteins and peripheral membrane proteins. This classification is paramount in understanding the diverse roles these proteins play within biological membranes. While both integral and peripheral membrane proteins are essential for the proper functioning of cells, they exhibit distinct structural and functional characteristics. In this section, we will delve into the contrasting attributes of integral and peripheral membrane proteins, shedding light on their roles, mechanisms of association with membranes, and key examples to illustrate these differences.

Integral Membrane Proteins: Embedded in the Membrane Matrix

Integral membrane proteins, as the name suggests, are deeply embedded within the lipid bilayer of cellular membranes. They traverse the hydrophobic core of the lipid bilayer, thereby becoming an integral part of the membrane structure. This intimate association with the lipid matrix is a defining feature of integral membrane proteins, and it necessitates specific structural adaptations to accommodate the hydrophobic environment.

Structural Features of Integral Membrane Proteins

One of the hallmark features of integral membrane proteins is their transmembrane domain (TMD), which typically consists of one or more hydrophobic α-helices or β-barrels. These hydrophobic segments interact with the lipid acyl chains, anchoring the protein securely within the membrane. Depending on their structure, integral membrane proteins can be further classified into single-pass and multi-pass proteins.

Single-Pass Integral Membrane Proteins: These proteins possess a single TMD that crosses the lipid bilayer once. An exemplary single-pass integral membrane protein is the glucose transporter GLUT1. This protein facilitates the transport of glucose across the plasma membrane, ensuring a steady supply of this essential energy source for the cell. GLUT1's single TMD is strategically positioned to enable glucose binding and transport.

Multi-Pass Integral Membrane Proteins: In contrast, multi-pass integral membrane proteins traverse the lipid bilayer multiple times, often forming α-helical bundles. The G protein-coupled receptors (GPCRs) exemplify this category. GPCRs are integral to cellular signal transduction, responding to a diverse array of ligands, including hormones and neurotransmitters. The seven transmembrane helices of GPCRs create a binding pocket within the membrane, allowing them to initiate intracellular signalling pathways upon ligand binding.

Functional Roles of Integral Membrane Proteins

Integral membrane proteins serve a plethora of essential functions in cells. Their deep integration within the membrane grants them direct access to the hydrophobic core of the bilayer, facilitating the transport of hydrophobic molecules, ions, and even water across membranes. Beyond transporters and receptors, integral membrane proteins include channels that selectively permit the passage of ions or molecules, such as the potassium ion channel Kv1.2, which regulates the flow of potassium ions across cell membranes, playing a pivotal role in maintaining cellular electrical excitability.

Furthermore, integral membrane proteins are integral to cellular adhesion and structural integrity. Integrins, for instance, are heterodimeric integral membrane proteins that link the extracellular matrix to the cytoskeleton, enabling cells to adhere, migrate, and signal in response to their external environment. Disruptions in integrin function are associated with various pathological conditions, including cancer metastasis and immune disorders.

Peripheral Membrane Proteins: Loosely Associated with the Membrane Surface

While integral membrane proteins reside deeply within the lipid bilayer, peripheral membrane proteins, in contrast, do not penetrate the hydrophobic core but are loosely associated with the membrane's surface. These proteins interact with the membrane primarily through electrostatic and non-covalent interactions, such as hydrogen bonding and van der Waals forces. Peripheral membrane proteins can be found on the inner or outer leaflet of the lipid bilayer, and they often serve as regulators of integral membrane proteins or participate in signal transduction pathways.

Structural Features of Peripheral Membrane Proteins

Peripheral membrane proteins lack the transmembrane domains characteristic of integral membrane proteins. Instead, they typically possess hydrophilic regions or domains that facilitate their reversible binding to membrane surfaces. One example of a peripheral membrane protein is the protein kinase C (PKC), which plays a central role in signal transduction pathways. PKC contains C1 and C2 domains that bind to membrane lipids in response to cellular signals, enabling its activation and participation in downstream signalling events.

Functional Roles of Peripheral Membrane Proteins

Peripheral membrane proteins exhibit diverse functions in cellular processes. Many of them are involved in cellular signalling, where they act as intermediaries between extracellular signals and intracellular responses. For instance, the Ras family of GTPases, including H-Ras and K-Ras, are peripheral membrane proteins that anchor to the inner leaflet of the plasma membrane. Upon activation, these proteins relay signals that regulate cell proliferation, differentiation, and survival.

Other peripheral membrane proteins are implicated in membrane trafficking and vesicle fusion events. The clathrin adaptor protein complex, AP-2, is one such example. AP-2 recruits cargo molecules at the plasma membrane, facilitating their incorporation into clathrin-coated vesicles for endocytosis, a crucial process in maintaining membrane homeostasis.

Key Differences Between Integral and Peripheral Membrane Proteins

The distinction between integral and peripheral membrane proteins extends beyond their structural features and functional roles. Several key differences highlight their unique characteristics:

Membrane Association: Integral membrane proteins are embedded within the lipid bilayer, whereas peripheral membrane proteins are associated with the membrane's surface.

Transmembrane Domains: Integral membrane proteins possess one or more transmembrane domains, while peripheral membrane proteins lack such domains.

Binding Strength: Integral membrane proteins have strong, hydrophobic interactions with the lipid bilayer, making their association stable, whereas peripheral membrane proteins have weaker, reversible interactions.

Function: Integral membrane proteins often function as transporters, receptors, or channels involved in molecule and ion transport, whereas peripheral membrane proteins frequently act as regulators or signalling intermediates.

Integral and peripheral membrane proteins are two distinct classes of proteins crucial for the proper functioning of biological membranes. Integral membrane proteins are deeply embedded within the lipid bilayer, with transmembrane domains that allow them to interact intimately with the hydrophobic environment. They play roles in transport, signalling, and structural integrity. In contrast, peripheral membrane proteins are loosely associated with the membrane's surface, interacting primarily through non-covalent interactions. They serve as regulators, mediators of signalling pathways, and facilitators of membrane trafficking events. Understanding the differences between these two classes of membrane proteins is essential for unravelling the complexity of cellular processes and the mechanisms by which cells interact with their environment.

2.3 Single-Pass vs. Multi-Pass Membrane Proteins

Membrane proteins constitute a diverse group of biomolecules that play crucial roles in a wide array of cellular functions. Their classification is primarily based on their relationship with the lipid bilayer, which forms the structural foundation of biological membranes. One important aspect of this classification is the differentiation between single-pass and multi-pass membrane proteins. This distinction is fundamental to understanding the structural and functional diversity within this protein class.

Definition and Basic Characteristics

Single-pass and multi-pass membrane proteins, as the names suggest, are defined by the number of times they traverse the lipid bilayer. These characteristics have profound implications for their overall structure, topology, and function.

Single-Pass Membrane Proteins: Single-pass membrane proteins, also known as monotopic membrane proteins, span the lipid bilayer only once. They are typically anchored to the membrane by a single hydrophobic transmembrane (TM) segment. The rest of the protein, including its catalytic or functional domains, resides either in the cytoplasm or extracellular space, depending on the protein's orientation.

Multi-Pass Membrane Proteins: In contrast, multi-pass membrane proteins traverse the lipid bilayer multiple times, with two or more hydrophobic TM segments. These segments act as anchor points, and the regions between them may fold into domains that extend into both the cytoplasm and extracellular space, creating a complex, three-dimensional structure that spans the membrane.

Examples of Single-Pass Membrane Proteins

To illustrate this classification further, let's delve into examples of single-pass membrane proteins and examine their diverse functions and structures.

Glycophorin A (GPA): Glycophorin A is an archetypal single-pass membrane protein found in red blood cells. It serves as a receptor for the malarial parasite Plasmodium falciparum. GPA's single TM segment anchors it to the lipid bilayer, while its extracellular region contains glycosylated residues that are important for its function as a binding partner for the parasite.

Cytochrome b5: Cytochrome b5 is another example of a single-pass membrane protein, which plays a critical role in electron transport within the endoplasmic reticulum. Its single TM domain embeds in the membrane, while its heme-binding domain extends into the cytoplasm. This topology allows cytochrome b5 to participate in redox reactions involving lipid-bound substrates.

Rhodopsin: Rhodopsin, a G-protein coupled receptor (GPCR), is yet another notable single-pass membrane protein. Its single TM helix anchors it in the lipid bilayer, while its extracellular domain houses a light-absorbing chromophore, essential for its function in phototransduction in rod cells of the retina.

These examples demonstrate that single-pass membrane proteins can serve diverse functions, including receptor-mediated signalling, electron transport, and sensory perception. Their relatively simple topology, with one TM domain, makes them accessible targets for structural studies using techniques like X-ray crystallography and NMR spectroscopy.

Examples of Multi-Pass Membrane Proteins

Multi-pass membrane proteins, on the other hand, exhibit more complex topologies and functional diversity. Here are some notable examples:

***G protein-coupled receptors (GPCRs)*:** GPCRs are multi-pass membrane proteins that are integral to signal transduction across the cell membrane. They typically consist of seven TM helices, forming a distinctive seven-transmembrane domain architecture. These helices create a binding pocket for ligands on the extracellular side and interact with intracellular signalling proteins on the cytoplasmic side, allowing them to relay signals from the extracellular environment to the interior of the cell.

***Adenosine triphosphate (ATP) synthase*:** ATP synthase is a multi-pass membrane protein complex found in the inner mitochondrial membrane. It is responsible for synthesizing ATP, the cell's primary energy currency, during oxidative phosphorylation. The complex comprises multiple TM segments, including a rotor subunit that spans the membrane and a catalytic domain on the matrix side of the membrane.

***Amyloid precursor protein (APP)*:** APP is another multi-pass membrane protein known for its role in Alzheimer's disease. It spans the membrane multiple times, with various proteolytic cleavage sites distributed across its structure. The proteolysis of APP can lead to the production of amyloid-beta peptides, which are implicated in the pathogenesis of Alzheimer's disease.

These examples illustrate the structural complexity and functional versatility of multi-pass membrane proteins. The presence of multiple TM segments allows them to interact with a diverse array of ligands, partners, and substrates both inside and outside the cell membrane.

Importance and Implications

Understanding the classification of single-pass and multi-pass membrane proteins is pivotal for several reasons:

Functional Implications: The number of TM segments significantly influences a membrane protein's function. Single-pass proteins often act as receptors, transporters, or anchors, while multi-pass proteins have more intricate functions, including signal transduction and enzymatic activities.

Structural Challenges: Determining the structure of multi-pass membrane proteins can be considerably more challenging due to their complex topology. Advanced techniques such as cryo-electron microscopy have been instrumental in elucidating their structures.

Pharmacological Relevance: Many drugs target membrane proteins, especially GPCRs. Understanding their multi-pass nature is crucial for designing drugs that modulate their activity selectively.

The distinction between single-pass and multi-pass membrane proteins is fundamental to grasping the structural and functional diversity within this class of biomolecules. The examples provided underscore how these proteins are involved in a wide range of cellular processes, making them pivotal subjects of study in structural biology and pharmacology.

As we delve further into the structural biology of membrane proteins in subsequent chapters, we will explore in greater detail the techniques and methodologies employed to study these proteins and unravel their intricate roles in cellular function and disease.

2.4 Glycoproteins in the Membrane

Glycoproteins represent a captivating class of membrane proteins, characterized by their intricate structural composition and pivotal functional roles. These biomolecules are integral to a multitude of cellular processes, ranging from cell signalling to immune response regulation. In this section, we delve into the fascinating world of glycoproteins embedded within cellular membranes, exploring their structural intricacies, functions, and significance in both health and disease.

Structural Characteristics of Glycoproteins

Glycoproteins are a diverse group of membrane proteins that are covalently linked to carbohydrate moieties. These carbohydrates, or glycans, can be bound to the protein backbone through N-glycosidic or O-glycosidic bonds, forming glycopeptides or glycolipids, respectively. The addition of glycans to proteins occurs in the endoplasmic reticulum and Golgi apparatus, and it is a highly regulated and precise process.

The structural diversity of glycoproteins arises from variations in glycan composition, length, and branching, as well as the site-specificity of glycosylation. These differences contribute to the functional diversity of glycoproteins, allowing them to participate in various biological processes.

Glycoproteins are classified into several categories based on their glycan structure and location within the cell. For instance, cell surface glycoproteins are often involved in cell adhesion, signalling, and immune response regulation, while glycoproteins found in the extracellular matrix contribute to tissue integrity and function.

Functions of Glycoproteins in Membranes

Glycoproteins play indispensable roles in various cellular functions, and their presence on the cell membrane is particularly noteworthy. Here, we explore some key functions of glycoproteins in the context of cell membranes:

Cell-Cell Recognition and Adhesion

One of the primary functions of glycoproteins on the cell membrane is to facilitate cell-cell recognition and adhesion. These glycoproteins often act as receptors, binding to specific ligands on neighbouring cells. For example, selectins are a family of cell adhesion molecules that contain lectin-like domains and are essential for leukocyte rolling during inflammation. The interaction between selectins and their ligands, which are glycoproteins bearing specific carbohydrate structures, enables leukocytes to adhere to endothelial cells and subsequently migrate to sites of infection or tissue damage.

Immune Response

Glycoproteins on the cell membrane also play a crucial role in the immune response. Major histocompatibility complex (MHC) proteins, including MHC class I and II, are glycoproteins that present antigens to immune cells. MHC class I molecules display endogenous antigens to cytotoxic T cells, while MHC class II molecules present antigens from extracellular pathogens to helper T cells. These interactions are pivotal for immune surveillance and the activation of immune responses.

Signal Transduction

Glycoproteins are integral components of many signalling pathways. Receptor tyrosine kinases (RTKs) are a class of glycoproteins that become activated upon ligand binding. For instance, the epidermal growth factor receptor (EGFR), a well-studied RTK, undergoes autophosphorylation upon binding to its ligand, triggering a cascade of intracellular signalling events that regulate cell growth and differentiation.

Structural Integrity and Barrier Function

Glycoproteins, along with glycolipids, contribute to the structural integrity and barrier function of cell membranes. The glycan structures on these molecules create a protective glycocalyx on the cell surface, which helps shield cells from mechanical and chemical damage. Additionally, glycoproteins play a role in membrane fluidity and stability, influencing the overall properties of the lipid bilayer.

Disease Implications of Glycoproteins

Glycoproteins are not only vital for normal cellular function but also implicated in various diseases when their structure or function is compromised. Here, we explore a few examples:

Cancer

Aberrant glycosylation patterns of cell surface glycoproteins are a hallmark of cancer. Altered glycosylation can affect cell adhesion, invasion, and immune evasion, promoting tumour progression and metastasis. For example, changes in the glycosylation of mucin glycoproteins are associated with the aggressiveness of certain cancers, including pancreatic and breast cancer.

Genetic Disorders

Deficiencies in enzymes responsible for glycosylation can lead to genetic disorders known as congenital disorders of glycosylation (CDG). These disorders can result in a wide range of symptoms, including developmental delays, intellectual disabilities, and organ dysfunction.

Viral Infections

Many viruses exploit host glycoproteins for entry into cells. The spike proteins of enveloped viruses, such as the SARS-CoV-2 virus responsible for COVID-19, contain glycosylation sites that modulate viral infectivity and immune recognition. Understanding these glycoprotein interactions is crucial for the development of antiviral therapies and vaccines.

Structural Studies of Glycoproteins in Membranes

Studying the structure of glycoproteins within cell membranes presents unique challenges due to the complexity of glycan structures and their heterogeneity. However, recent advances in structural biology techniques have enabled researchers to unravel the intricacies of glycoprotein structures.

Cryo-Electron Microscopy (Cryo-EM)

Cryo-EM has emerged as a powerful tool for studying the structures of glycoproteins in their native membrane environment. By freezing samples at cryogenic temperatures, researchers can preserve the structural integrity of glycoproteins, allowing for high-resolution imaging of these molecules within the lipid bilayer.

NMR Spectroscopy

Nuclear magnetic resonance (NMR) spectroscopy is another technique used to investigate the structures and dynamics of glycoproteins in membranes. Isotope labelling and advanced NMR methodologies enable researchers to obtain valuable insights into the conformation and interactions of glycan chains with the protein backbone.

X-ray Crystallography

While challenging, X-ray crystallography has been employed to determine the structures of glycoproteins, particularly for smaller, soluble glycoproteins or glycoprotein domains. This technique has provided valuable insights into the arrangement of glycans and their interactions with the protein core.

Future Directions in Glycoprotein Research

The study of glycoproteins in membranes is a rapidly evolving field with numerous avenues for future exploration:

Glycoengineering

Advances in glycoengineering techniques may allow for the manipulation of glycan structures on glycoproteins. This could lead to the development of novel therapeutics, vaccines, and targeted drug delivery systems.

Integrating Structural and Functional Studies

Further integration of structural and functional studies will provide a more comprehensive understanding of glycoproteins' roles in health and disease. Combining techniques like cryo-EM, NMR, and X-ray crystallography with functional assays will shed light on the dynamic nature of glycoproteins.

Drug Discovery

Glycoproteins have emerged as attractive targets for drug discovery. Investigating the structural details of glycoprotein-ligand interactions can aid in the design of novel drugs that modulate glycoprotein function.

Glycoproteins in cell membranes are multifaceted molecules with diverse functions that are crucial for cellular processes. Their structural complexity and functional significance make them a fascinating subject of study in the field of membrane protein structural biology. Advancements in technology and techniques continue to unravel the mysteries of these glycosylated membrane proteins, opening up new possibilities for therapeutic interventions and a deeper understanding of health and disease.

Chapter 3: Membrane Protein Isolation and Purification

3.1 Techniques for Membrane Protein Isolation

Membrane proteins, often referred to as the gatekeepers of the cell, play a pivotal role in a wide array of biological processes, from ion transport and signal transduction to cell adhesion and molecular recognition. Their diverse functions make them prime targets for structural and functional studies, but the isolation and purification of these proteins present unique challenges due to their inherent hydrophobicity and integration within lipid bilayers. In this section, we delve into the techniques employed for the isolation of membrane proteins, exploring both classical and modern approaches, and shedding light on their strengths and limitations.

Detergent-Based Solubilization

One of the foundational techniques for isolating membrane proteins involves the use of detergents, amphipathic molecules that possess both hydrophobic and hydrophilic regions. Detergents disrupt the lipid bilayer by inserting their hydrophobic tails into the lipid acyl chains, thereby solubilizing the membrane protein complexes within. This method is often the initial step in the isolation process.

Solubilization Strategies: Various detergents are employed in membrane protein solubilization, each with distinct properties and applications. Non-ionic detergents, such as Triton X-100 and octylglucoside, are milder and generally used for initial solubilization. In contrast, zwitterionic detergents, like dodecyl maltoside, have gained popularity for their ability to maintain protein stability during solubilization. Anionic and cationic detergents, such as sodium dodecyl sulphate (SDS) and cetyltrimethylammonium bromide (CTAB), are less commonly used due to their potential to denature proteins, but they find application in specific cases.

Challenges in Detergent-Based Solubilization: While detergent-based solubilization is a common technique, it has its limitations. The choice of detergent can significantly impact the stability and functionality of the target membrane protein. Detergents may disrupt protein-protein interactions, alter the native conformation, or even lead to protein denaturation. Moreover, excessive detergent concentrations can form mixed micelles, rendering the protein unsuitable for downstream applications.

Membrane Extraction with Chaotropic Agents

In some cases, detergents may not suffice for efficient membrane protein isolation, especially for integral membrane proteins deeply embedded in lipid bilayers. Chaotropic agents, such as urea and guanidine hydrochloride, are used to disrupt lipid-protein interactions by denaturing the lipid bilayer structure.

Chaotropic Agents in Membrane Protein Extraction: Urea and guanidine hydrochloride are strong chaotropic agents that disrupt hydrogen bonds and hydrophobic interactions, thus solubilizing membrane proteins. They are especially useful for extracting integral membrane proteins or when detergents fail to efficiently solubilize the target proteins.

Limitations and Considerations: While chaotropic agents can be effective, they pose risks of protein denaturation and loss of native structure. Moreover, their use often necessitates subsequent refolding steps to restore the protein to its native state, which can be a challenging and time-consuming process.

Membrane Protein Extraction by Phase Separation

Phase separation techniques leverage the immiscibility of two phases—usually an aqueous phase and an organic phase—to selectively partition membrane proteins from the lipid bilayer. One of the most widely used phase separation methods is the aqueous two-phase system (ATPS).

Aqueous Two-Phase Systems: ATPS involves the creation of two immiscible phases in a single solution, typically composed of polymers and salts. Membrane proteins preferentially partition into one of the phases based on their hydrophobicity. This technique has the advantage of yielding a gentle and biocompatible environment for membrane proteins.

Benefits and Applications: ATPS is particularly useful for preserving the native structure and function of membrane proteins, making it an ideal choice when studying their biological activity. It has been employed successfully in the isolation of membrane transporters and receptors, among other classes of membrane proteins.

Immunoaffinity Chromatography

Immunoaffinity chromatography is a highly specific technique that exploits the binding affinity between antibodies and antigens to isolate membrane proteins selectively. This method is particularly valuable for purifying specific membrane protein isoforms or complexes.

Procedure: In immunoaffinity chromatography, antibodies specific to the target membrane protein are immobilized on a solid support, such as a chromatography column or magnetic beads. The sample containing the membrane protein mixture is then passed through the column, where the target protein binds to the immobilized antibodies. After washing away unbound proteins, the target protein is eluted, providing a highly purified sample.

Applications and Considerations: Immunoaffinity chromatography is an indispensable tool in the isolation of membrane proteins with specific post-translational modifications or protein-protein interactions. However, it requires prior knowledge of the target protein and access to high-quality antibodies, which may limit its application in certain cases.

Biotinylation and Streptavidin-Based Purification

Biotinylation involves the covalent attachment of biotin molecules to specific amino acid residues on a membrane protein. Streptavidin, a protein with a strong affinity for biotin, is then used to selectively capture and purify the biotinylated membrane protein.

Procedure: To utilize this technique, the target membrane protein is genetically engineered to incorporate a biotin acceptor peptide (BAP) tag. This tag allows for site-specific biotinylation using the enzyme biotin ligase. The biotinylated protein is subsequently captured using streptavidin-coated affinity resins or magnetic beads.

Advantages and Applications: Biotinylation and streptavidin-based purification offer high specificity and minimal interference with the native structure and function of the target protein. This technique has found extensive use in studying the stoichiometry of membrane protein complexes and their interactions with other biomolecules.

The isolation and purification of membrane proteins represent a critical initial step in structural and functional studies. Researchers have at their disposal a diverse array of techniques, each with its own advantages and limitations. The choice of method depends on factors such as the hydrophobicity of the target protein, its specific interactions, and the downstream applications. Detergent-based solubilization remains a common starting point, while phase separation techniques, immunoaffinity chromatography, and innovative approaches like biotinylation continue to advance the field, enabling researchers to unlock the mysteries of these vital cellular components. In the subsequent sections of this chapter, we will explore strategies for the purification and characterization of membrane proteins once they have been successfully isolated.

3.2 Detergents and Their Role in Solubilization

In structural biology, the isolation and purification of membrane proteins are formidable tasks. Membrane proteins, as their name suggests, reside within lipid bilayers, making them inherently hydrophobic. This hydrophobicity poses a significant challenge when attempting to extract these proteins from their native environment and maintain their structural integrity. Enter detergents—remarkable amphiphilic molecules that play a pivotal role in solubilizing membrane proteins, rendering them amenable to further analysis. In this section, we will delve into the world of detergents, exploring their diverse properties and their indispensable role in the study of membrane proteins.

The Hydrophobic Dilemma

Before we delve into the world of detergents, it's crucial to grasp the inherent hydrophobicity of membrane proteins. These proteins, which span the cellular membrane, consist of hydrophobic transmembrane domains that are typically composed of nonpolar amino acids. This hydrophobic core is surrounded by hydrophilic regions that interact with the aqueous environment inside and outside the cell. The stark contrast between these hydrophilic and hydrophobic domains presents a formidable challenge when attempting to isolate and purify membrane proteins.

Imagine trying to extract a protein that is innately repelled by water from a watery milieu—akin to attempting to pluck a bead of oil from a pool of water. This is where detergents come into play, acting as molecular intermediaries that bridge the divide between the hydrophobic protein core and the aqueous environment.

Detergents: Amphiphilic Saviors

Detergents are amphiphilic molecules, meaning they possess both hydrophilic (water-attracting) and hydrophobic (water-repelling) regions within the same molecule. This unique dual nature allows detergents to encapsulate hydrophobic portions of membrane proteins, effectively forming micelles or lipoprotein particles. In essence, detergents provide a molecular shield, enveloping the hydrophobic domains of membrane proteins and creating a stable aqueous environment in which these proteins can reside. Let's explore the properties and types of detergents commonly used in membrane protein research.

Properties of Detergents

Critical Micelle Concentration (CMC): Each detergent has a CMC, a concentration at which it spontaneously forms micelles. Below the CMC, detergent molecules are dispersed in solution; above it, they aggregate into micelles. The CMC is a critical parameter, as it influences the efficiency of solubilization.

Hydrophilic Headgroup: The hydrophilic portion of a detergent is typically a charged or polar group. Common headgroups include sulphate, carboxylate, ammonium, and glycoside moieties. These headgroups interact favourably with water molecules.

Hydrophobic Tail: The hydrophobic tail of a detergent consists of a long hydrocarbon chain, which provides the molecule with its lipophilic properties. The length and saturation of the hydrocarbon tail can vary among detergents, influencing their solubilization capabilities.

Charge: Detergents can be categorized into three groups based on charge: anionic (negatively charged), cationic (positively charged), and non-ionic (neutral). The choice of detergent often depends on the specific properties of the target membrane protein.

Types of Detergents

There is a diverse array of detergents available to researchers, each with distinct properties and suitability for different applications. Here are a few examples:

Sodium Dodecyl Sulphate (SDS): A classic anionic detergent, SDS is widely used in electrophoresis to denature and solubilize proteins. Its strong negative charge imparts a uniform charge-to-mass ratio to proteins, facilitating their separation.

Triton X-100: A non-ionic detergent, Triton X-100 is frequently employed for cell lysis and protein extraction. Its relatively mild properties make it suitable for maintaining protein structure in some applications.

N-Octyl-β-D-Glucopyranoside (OG): This non-ionic detergent is commonly used in membrane protein crystallization and structural studies due to its ability to maintain protein stability in lipidic environments.

Dodecylphosphocholine (DPC): A zwitterionic detergent, DPC is known for its compatibility with nuclear magnetic resonance (NMR) spectroscopy, making it a preferred choice for studying membrane protein structures in solution.

Choosing the Right Detergent

Selecting the appropriate detergent is a critical step in membrane protein solubilization. The choice depends on various factors, including the protein's hydrophobicity, stability, and downstream applications. It's essential to strike a balance between effective solubilization and maintaining the protein's native conformation.

For example, when working with highly hydrophobic membrane proteins, detergents with long hydrophobic tails may be necessary to solubilize them effectively. However, these detergents can also disrupt protein stability, necessitating the addition of stabilizing agents.

Challenges and Considerations

While detergents have revolutionized membrane protein research, they are not without challenges and limitations. Some of these include:

Detergent-Induced Protein Denaturation: Detergents can disrupt protein-protein interactions and even denature proteins if used in excess. Careful optimization of detergent concentration is crucial to maintain protein integrity.

Detergent Exchange: Detergents used during purification must be removed before subsequent structural analysis. This detergent exchange step can be technically demanding and may impact protein stability.

Detergent Effects on Membrane Mimetics: In studies aiming to reconstitute membrane proteins in artificial lipid bilayers or nanodiscs, the choice of detergent can significantly affect membrane mimetic properties, potentially influencing protein behaviour.

Detergent Toxicity: Some detergents can be toxic to cells, limiting their use in expression systems. Researchers must consider potential toxicity and explore alternatives when working with live cells.

Considering the significance of membrane protein isolation and purification, detergents are the silent choreographers that allow researchers to bring these hydrophobic proteins onto the stage of structural biology. Their amphiphilic nature, along with their various properties and types, makes detergents versatile tools for solubilizing membrane proteins, paving the way for subsequent structural and functional studies. However, the judicious choice and careful handling of detergents are paramount to preserving the native conformation and biological relevance of these vital cellular components. As the field of membrane protein research continues to advance, so too will our understanding of the intricate roles these proteins play in cellular life.

3.3 Challenges in Membrane Protein Purification

The isolation and purification of membrane proteins represent a formidable challenge in structural biology. Membrane proteins, which play vital roles in cellular processes, are embedded within lipid bilayers, making them inherently hydrophobic. This hydrophobic nature not only poses difficulties in extracting them from the native membrane environment but also introduces various challenges during the purification process. In this section, we will delve into the intricacies of membrane protein purification, highlighting the major challenges faced by researchers in this field.

Hydrophobicity and Solubilization

One of the primary obstacles in membrane protein purification arises from their hydrophobicity. The hydrophobic interactions between membrane proteins and the lipid bilayer are essential for their structural integrity and biological function. However, these interactions also make it challenging to solubilize membrane proteins in aqueous solutions. The lipid bilayer provides a stable hydrophobic environment, and disrupting this environment without denaturing the protein is a complex task.

Researchers commonly employ detergents to solubilize membrane proteins. Detergents possess both hydrophobic and hydrophilic regions, allowing them to interact with the hydrophobic transmembrane domains of membrane proteins while keeping them in a soluble state. However, the choice of detergent is critical, as different detergents can have varying effects on protein stability and activity. Moreover, excessive use of detergents can lead to protein denaturation or aggregation, further complicating the purification process.

Native Conformation Preservation

Preserving the native conformation of membrane proteins during purification is paramount for subsequent structural and functional studies. Many membrane proteins are sensitive to changes in their environment, and deviations from their native conditions can result in structural alterations or loss of function.

During the purification process, researchers must carefully control parameters such as pH, temperature, and ionic strength to maintain the stability of membrane proteins. Additionally, the use of mild detergents and lipid mimetics is crucial to mimic the native lipid bilayer environment and ensure that the protein retains its native structure.

One example of successful native conformation preservation is the purification of bacteriorhodopsin, a light-driven proton pump found in the cell membrane of Halobacterium salinarum. By using a combination of detergents and lipids that mimic the native lipid environment, researchers were able to obtain bacteriorhodopsin crystals that closely resembled its native structure, allowing for high-resolution structural determination.

Heterogeneity and Post-Translational Modifications

Another challenge in membrane protein purification is the presence of heterogeneity within the protein sample. Membrane proteins often undergo various post-translational modifications (PTMs), such as glycosylation, phosphorylation, and lipidation, which can introduce heterogeneity into the protein population.

Glycosylation, in particular, is a common PTM in membrane proteins. It involves the addition of carbohydrate moieties to specific amino acid residues and can have a significant impact on protein structure and function. Purifying glycosylated membrane proteins while maintaining the diversity of glycoforms is a complex task.

One example of a glycosylated membrane protein is the epidermal growth factor receptor (EGFR). EGFR is a receptor tyrosine kinase that plays a crucial role in cell signalling. Its purification involves specific strategies to preserve the glycosylation patterns, as alterations in glycosylation can affect its ligand-binding affinity and downstream signalling.

Membrane Protein Stability

The stability of membrane proteins can vary widely, with some proteins being more robust than others. Maintaining the stability of fragile membrane proteins during the purification process is challenging, as they can be prone to unfolding, aggregation, or degradation.

To address this challenge, researchers often employ strategies such as co-purification with stabilizing agents or chaperone proteins. For instance, the bacterial outer membrane protein OmpA has been successfully purified and stabilized using maltose-binding protein (MBP) as a fusion partner. MBP not only aids in solubilizing OmpA but also prevents its aggregation, enabling its subsequent structural studies.

Low Expression Levels

In many cases, membrane proteins are expressed at low levels in their native hosts, which can further complicate the purification process. Low expression levels make it challenging to obtain sufficient quantities of protein for structural and functional studies.

To overcome this challenge, researchers often turn to heterologous expression systems, such as Escherichia coli, yeast, or insect cells, which can yield higher expression levels of membrane proteins. However, the choice of expression system must be carefully considered, as it can impact protein folding, post-translational modifications, and functionality.

For example, the aquaporin water channels, which are integral membrane proteins, were successfully overexpressed in E. coli using a fusion protein strategy. This approach not only increased the expression levels but also facilitated their purification and subsequent structural analysis.

Detergent Removal and Lipid Reconstitution

Once membrane proteins are solubilized and purified, it is often necessary to remove detergents used in the solubilization step and reconstitute the protein into a lipid bilayer environment. Detergent removal is critical to prevent interference with downstream assays or structural studies. However, the process of detergent removal can be challenging and may lead to protein instability or aggregation.

Lipid reconstitution, on the other hand, aims to mimic the native lipid environment of the membrane protein. This step is essential for maintaining the native conformation and function of the protein. Careful selection of lipid composition and reconstitution methods is required to ensure the protein's proper insertion into the lipid bilayer.

Membrane protein purification is a complex and multifaceted process that involves overcoming various challenges related to hydrophobicity, native conformation preservation, heterogeneity, stability, low expression levels, detergent removal, and lipid reconstitution. Successful purification strategies often require a combination of biochemical, biophysical, and structural biology techniques tailored to the specific characteristics of the target protein. Researchers continue to develop innovative methods and technologies to address these challenges, driving advancements in our understanding of membrane protein structure and function.

3.4 Purification Strategies

In the world of membrane protein structural biology, the successful isolation and purification of these enigmatic molecules are pivotal milestones. Precise purification strategies are crucial not only for obtaining sufficient quantities of purified membrane proteins but also for preserving their structural and functional integrity. This section delves into various purification strategies employed by scientists to tackle the challenges presented by membrane proteins, offering insights and real-world examples.

Purification of membrane proteins from their native environment, the lipid bilayer, is a formidable task due to their hydrophobic nature and susceptibility to denaturation. Researchers have devised ingenious strategies over the years, often combining multiple approaches to overcome these hurdles. These strategies are not one-size-fits-all; they depend on the protein of interest, its expression system, and the downstream applications. In this section, we explore some widely used purification strategies and their practical applications.

Detergent-Based Solubilization

One of the primary challenges in membrane protein purification is extracting these hydrophobic entities from the lipid bilayer without compromising their structural integrity. Detergents have been the workhorse in this regard. These amphiphilic molecules possess a hydrophobic tail that interacts with the lipid bilayer while their hydrophilic head groups shield the hydrophobic regions, thereby solubilizing the protein. Detergent-based solubilization has been successfully employed in numerous membrane protein purification attempts.

A Classic Example: Bacteriorhodopsin

The purple membrane of Halobacterium salinarum contains a renowned membrane protein, bacteriorhodopsin, which functions as a light-driven proton pump. Early studies on bacteriorhodopsin purification by Oesterhelt and Stoeckenius (1973) exemplify the use of detergents. They employed the mild detergent octylglucoside to solubilize the protein while maintaining its functionality, paving the way for subsequent structural studies.

Size Exclusion Chromatography

Once membrane proteins are solubilized, it is imperative to separate them from lipids, nucleic acids, and other contaminants. Size exclusion chromatography (SEC), also known as gel filtration chromatography, is a reliable technique for this purpose. It exploits the differences in molecular size to separate molecules, allowing membrane proteins to be eluted in a distinct fraction based on their size.

A Notable Application: Rhodopsin

Rhodopsin, a G-protein-coupled receptor (GPCR) responsible for vision in mammals, was purified by Palczewski and colleagues (1994) using SEC. The protein's solubilized form was separated from lipids and other impurities by running it through a size exclusion column. The resulting purified rhodopsin was integral for elucidating its structure and function.

Affinity Chromatography

Affinity chromatography is a purification technique that capitalizes on the specific interactions between a target molecule and a ligand immobilized on a solid support matrix. Immobilizing ligands that bind selectively to the membrane protein of interest allows for its isolation with high purity and yield.

An Exemplary Case: Green Fluorescent Protein (GFP)-Tagged Receptors

In studies involving the purification of GPCRs, which are notorious for their low expression levels, GFP tagging combined with affinity chromatography has been highly successful. Researchers have fused GFP to the C-terminus of GPCRs, enabling the selective purification of the tagged receptor using a solid support matrix conjugated to an anti-GFP antibody. This method has been pivotal in the structural analysis of GPCRs.

Differential Centrifugation

Differential centrifugation is a fundamental technique in cell biology and membrane protein purification. It involves subjecting a cell lysate or membrane fraction to a series of centrifugation steps at increasing speeds. This separates cellular components based on their size and density.

Applying Differential Centrifugation to Mitochondrial Membrane Proteins

Mitochondria, the powerhouses of the cell, contain a plethora of membrane proteins. To isolate these proteins, researchers employ differential centrifugation by first homogenizing cells to release mitochondria and then subjecting the homogenate to sequential centrifugation steps. This method has paved the way for in-depth studies of mitochondrial membrane proteins and their role in cellular energetics.

Protein-Lipid Nanodiscs

As an alternative to traditional detergent solubilization, the use of lipid nanodiscs has gained popularity in recent years. Nanodiscs are small, disk-like structures composed of lipids and membrane scaffold proteins (MSPs) that encapsulate membrane proteins, providing a native-like lipid environment.

Nanodiscs in Action: Cytochrome P450 Enzymes

Cytochrome P450 enzymes play a pivotal role in drug metabolism and are notoriously challenging to study due to their hydrophobic nature. By incorporating cytochrome P450 enzymes into nanodiscs, researchers have created stable, functionally active complexes that retain their native-like lipid environment. This approach has opened new avenues for investigating drug interactions and metabolism.

Purifying membrane proteins remains an art as much as a science. The strategies outlined here represent just a fraction of the diverse methods available to researchers in this field. The choice of purification strategy depends on various factors, including the target protein, its expression system, and the intended downstream applications. As technology continues to advance, and our understanding of membrane protein biology deepens, we can anticipate even more innovative purification strategies on the horizon. The ability to successfully isolate and purify these proteins is paramount for elucidating their structures and functions, ultimately leading to breakthroughs in drug discovery and our understanding of fundamental cellular processes.

Chapter 4: Membrane Protein Expression Systems

4.1 Cell-Free Expression Systems

The expression of membrane proteins for structural studies is a formidable task. Traditional methods, such as bacterial and eukaryotic expression systems, often encounter difficulties due to the hydrophobic nature and complex topology of membrane proteins. In recent years, cell-free expression systems have emerged as a powerful alternative for membrane protein production. In this section, we delve into the world of cell-free expression systems, exploring their principles, advantages, and notable examples.

Principles of Cell-Free Expression

Cell-free expression systems, also known as in vitro protein synthesis systems, offer a unique approach to produce membrane proteins outside of living cells. Unlike conventional cellular systems, these systems operate in a test tube, bypassing many of the challenges associated with cellular expression. The fundamental principle behind cell-free expression is the use of cellular machinery, specifically the translational machinery, extracted from cells and reconstituted in a controlled environment. This machinery can be obtained from various sources, including bacterial, eukaryotic, or even cell-free extracts from natural sources like E. coli, wheat germ, or rabbit reticulocytes.

The core components of a cell-free expression system typically include ribosomes, tRNA molecules, amino acids, and energy sources (usually ATP, GTP, etc.). Importantly, the membrane protein's gene of interest is added to the system in the form of a DNA template, which is transcribed into messenger RNA (mRNA). The ribosomes then read the mRNA and synthesize the target protein by linking amino acids together, following the genetic code. This process mimics the natural protein synthesis that occurs within cells but in a controlled, cell-free environment.

Advantages of Cell-Free Expression Systems

Cell-free expression systems offer several advantages that make them particularly attractive for membrane protein production in structural biology:

Flexibility in Membrane Protein Sources

One of the key advantages of cell-free expression systems is their versatility in accommodating membrane proteins from various sources. Unlike cellular expression systems that are tailored to specific organisms, cell-free systems can work with proteins from prokaryotic, eukaryotic, or even archaeal origin. This flexibility allows researchers to express a wide range of membrane proteins, including those from extremophiles or organisms that are challenging to culture.

Enhanced Control over Expression

In cell-free systems, researchers have precise control over the reaction conditions, including temperature, pH, and the concentrations of key components. This level of control is particularly valuable when working with delicate or difficult-to-express membrane proteins. Researchers can optimize conditions to maximize protein yield and folding, thereby improving the chances of obtaining high-quality samples for structural studies.

Rapid Protein Production

Cell-free expression systems are known for their speed. They can produce membrane proteins in a matter of hours to days, compared to the days or weeks often required in cellular expression systems. This rapid turnaround time is especially advantageous for projects with tight timelines, such as drug discovery efforts targeting membrane proteins.

Isotope Labelling

For structural studies involving nuclear magnetic resonance (NMR) spectroscopy, isotope labelling is crucial. Cell-free systems allow for straightforward incorporation of isotopically labelled amino acids, simplifying the production of proteins suitable for NMR experiments. This feature is particularly important for elucidating the structure and dynamics of membrane proteins.

Notable Examples of Membrane Proteins Produced Using Cell-Free Systems

Numerous membrane proteins have been successfully expressed using cell-free expression systems, demonstrating the broad applicability of this approach. Here are a few notable examples:

Bacterial Transporters: EmrE

EmrE, a small multidrug transporter from Escherichia coli, was one of the early successes of cell-free expression for membrane proteins. Researchers used a cell-free system to produce sufficient quantities of EmrE, enabling structural studies by NMR spectroscopy. This work provided valuable insights into the protein's conformational changes during drug transport.

Eukaryotic GPCRs: β2-Adrenergic Receptor

The β2-adrenergic receptor, a prototypical G-protein-coupled receptor (GPCR) found in eukaryotes, is notoriously challenging to express in cellular systems due to its complex folding requirements. Cell-free expression systems have been instrumental in producing functional β2-adrenergic receptor samples for structural studies, contributing to our understanding of GPCR activation and signalling.

Human Ion Channels: Kv1.2 Potassium Channel

Kv1.2 is a voltage-gated potassium channel crucial for neuronal function. Cell-free expression systems have been employed to produce functional Kv1.2 channels, facilitating structural investigations using techniques like X-ray crystallography and cryo-electron microscopy. These studies have shed light on the channel's gating mechanism.

Viral Proteins: Influenza M2 Ion Channel

The influenza M2 ion channel is essential for the virus's replication and has been a target for antiviral drug development.

Cell-free expression systems have enabled the production of M2 channel protein for structural studies. This work has guided the design of antiviral drugs targeting the channel.

Challenges and Considerations

While cell-free expression systems offer many advantages, they are not without challenges and considerations:

Membrane Mimicry

To ensure proper folding and function, it's important to include membrane-mimicking components such as lipids or detergents in the cell-free reaction. Finding the right membrane environment that matches the native conditions of the protein can be a delicate balancing act.

Scale-Up

While cell-free systems are rapid, they may be less efficient at producing large quantities of protein compared to cellular systems. Researchers must carefully consider scale-up strategies if large quantities of membrane protein are needed.

Cost

The cost of reagents and specialized equipment for cell-free expression can be higher compared to traditional expression systems. Researchers should weigh the benefits against the costs for their specific project.

Cell-free expression systems have revolutionized the production of membrane proteins for structural biology. Their flexibility, control, and rapidity make them indispensable tools for researchers seeking to unravel the mysteries of membrane protein structure and function. Notable successes in producing a wide range of membrane proteins underscore the potential of cell-free systems in advancing our understanding of these crucial biomolecules.

4.2 Bacterial Expression Systems

Bacterial expression systems have revolutionized the field of structural biology, offering researchers a versatile and cost-effective means to produce membrane proteins for structural and functional studies. Among these systems, the **Escherichia coli** (E. coli) expression system stands out as the workhorse due to its simplicity and well-established protocols. In this section, we will delve into the intricacies of bacterial expression systems, highlighting their advantages, challenges, and providing notable examples of successful membrane protein expression using E. coli.

The Versatility of E. coli

Escherichia coli, a bacterium commonly found in the human gut, has been a staple in molecular biology research for decades. Its popularity as an expression host for membrane proteins is rooted in several key attributes:

Speed and Cost-Efficiency

E. coli's rapid growth rate and ease of manipulation make it an attractive choice for expressing membrane proteins. A typical E. coli culture can multiply every 20-30 minutes, allowing for large-scale protein production within a short time frame. Additionally, the cost of culturing E. coli is relatively low compared to eukaryotic expression systems, making it ideal for budget-conscious research projects.

Availability of Genetic Tools

Researchers have developed an extensive toolkit for genetic manipulation of E. coli, including plasmids for protein expression and various strains with specific genetic modifications. These tools enable precise control over gene expression, protein localization, and fusion protein construction, enhancing the flexibility of the system.

High Protein Yields

E. coli can produce high yields of recombinant membrane proteins, making it well-suited for structural biology studies. By optimizing growth conditions and induction protocols, researchers can achieve expression levels in the milligram to gram range, facilitating downstream purification and structural analysis.

Challenges in E. coli Expression of Membrane Proteins

While E. coli offers many advantages, it also presents unique challenges when expressing membrane proteins. Membrane proteins are inherently hydrophobic, making them prone to aggregation and misfolding in the aqueous cytoplasm of E. coli. Overcoming these challenges requires careful planning and optimization:

Protein Toxicity

Some membrane proteins, particularly those with toxic domains, can be detrimental to E. coli's growth and viability. To address this issue, researchers have developed strategies such as tightly regulated promoters and conditional expression systems to minimize the negative impact on host cells.

Misfolding and Aggregation

The hydrophobic nature of membrane proteins can lead to misfolding and aggregation. Co-expression with chaperones or fusion to solubility-enhancing tags like maltose-binding protein (MBP) or glutathione-S-transferase (GST) can mitigate these issues by promoting proper folding and preventing aggregation.

Membrane Insertion and Localization

Inserting membrane proteins into the lipid bilayer is a crucial step in their functional expression. E. coli's cytoplasmic membrane differs significantly from eukaryotic membranes, making it essential to consider the choice of membrane-targeting signals and fusion partners. Signal peptides, transmembrane domains, and signal-anchor sequences play critical roles in directing proteins to the appropriate membrane compartment.

Post-Translational Modifications

E. coli lacks the machinery for many eukaryotic post-translational modifications (PTMs), such as glycosylation and disulfide bond formation. This limitation can be a drawback when studying membrane proteins that rely on specific PTMs for stability or function. Alternative expression hosts or in vitro systems may be necessary for such cases.

Notable Success Stories in E. coli Expression

Despite the challenges, numerous membrane proteins have been successfully expressed in E. coli for structural and functional studies. These achievements highlight the adaptability of the system and the creative strategies employed by researchers:

Bacterial Outer Membrane Proteins

Bacterial outer membrane proteins, like OmpC and OmpF, serve as classic examples of successful E. coli expression. Their relatively simple structure and localization in the outer membrane make them amenable to expression in their native host. Researchers have exploited the natural targeting mechanisms of E. coli to achieve high yields of these proteins.

G-Protein Coupled Receptors (GPCRs)

GPCRs are a challenging class of membrane proteins due to their complex folding and post-translational modifications. However, recent advances in E. coli expression have led to the successful production of several GPCRs, including the β2-adrenergic receptor, by optimizing expression conditions and using fusion partners that promote proper folding.

Transporters and Channels

Membrane transporters and ion channels, like the lactose permease LacY and the potassium channel KcsA, have been expressed and purified from E. coli with great success. Rational design of fusion constructs, such as fusions with green fluorescent protein (GFP), has facilitated their structural and functional studies.

Antibiotic Resistance Proteins

Bacterial membrane proteins involved in antibiotic resistance, such as the AcrB efflux pump, have been expressed in E. coli to understand their mechanisms of action. These studies have provided valuable insights into drug resistance mechanisms.

Advances in E. coli Expression Strategies

Continual improvements in E. coli expression systems have expanded their utility for membrane protein research. Recent advances include the development of novel fusion partners, alternative host strains, and custom-made synthetic biology tools. Additionally, innovations in protein purification and detergent selection have enhanced the quality of membrane protein samples.

Bacterial expression systems, particularly the E. coli system, have played a pivotal role in advancing our understanding of membrane proteins. Their versatility, cost-effectiveness, and success stories in expressing diverse membrane proteins have made them indispensable tools in structural biology. Despite the challenges, ongoing research continues to refine and expand the capabilities of these systems, promising exciting discoveries in the field of membrane protein structural biology.

4.3 Eukaryotic Expression Systems

In the quest to unravel the intricate structures and functions of membrane proteins, researchers often find themselves at the crossroads of choosing the most suitable expression system for their studies. While prokaryotic expression systems like E. coli have historically been valuable for the heterologous expression of soluble proteins, the complexity and diversity of eukaryotic membrane proteins often necessitate the use of eukaryotic expression systems. In this section, we explore the advantages, challenges, and notable examples of employing eukaryotic expression systems for membrane protein production.

The Advantages of Eukaryotic Expression Systems

Eukaryotic expression systems, primarily represented by yeast, insect cells, and mammalian cells, offer several distinct advantages for the production of membrane proteins. These advantages, stemming from the closer resemblance to the native cellular environment, play a pivotal role in obtaining correctly folded and biologically functional membrane proteins.

Proper Protein Folding and Post-Translational Modifications: Eukaryotic cells provide a more native-like environment for membrane protein expression. They possess the machinery for extensive post-translational modifications (PTMs), including glycosylation, phosphorylation, and disulfide bond formation, which are crucial for the proper folding and function of many membrane proteins. This capability is particularly relevant for G-protein coupled receptors (GPCRs) and ion channels, which often require specific PTMs for functional integrity.

Membrane Localization: Eukaryotic expression systems enable the proper targeting of membrane proteins to the endoplasmic reticulum (ER) and other subcellular compartments, facilitating their insertion into lipid bilayers. This is critical for multi-pass transmembrane proteins, as they must correctly span the lipid bilayer to perform their biological functions.

Formation of Complexes: Many membrane proteins function as part of larger complexes. Eukaryotic systems allow the co-expression of interacting partners, enabling the formation of these complexes in their natural context. For example, the study of the γ-secretase complex, involved in Alzheimer's disease, necessitated the co-expression of multiple subunits to recapitulate its functional state.

Challenges in Eukaryotic Expression Systems

While eukaryotic expression systems offer undeniable advantages, they also come with their set of challenges that researchers must address to successfully produce and study membrane proteins.

Complexity and Cost: Eukaryotic expression systems are inherently more complex and expensive than prokaryotic systems. Mammalian cell culture, in particular, requires specialized equipment, culture media, and a controlled environment, significantly increasing the overall cost of experiments.

Low Expression Levels: Achieving high levels of membrane protein expression in eukaryotic systems can be challenging. The regulation of gene expression is tighter in eukaryotes to avoid cellular stress, often leading to lower protein yields compared to prokaryotic systems.

Cell Line Specificity: Different membrane proteins may require different eukaryotic cell lines for optimal expression. Choosing the appropriate cell line can be a time-consuming process, and sometimes multiple cell lines need to be tested to identify the most suitable one.

Protein Stability and Toxicity: Overexpression of membrane proteins, especially those with toxic effects on the host cell, can lead to poor cell viability and protein degradation. Strategies to mitigate toxicity, such as using inducible promoters or co-expression with chaperones, must be employed.

Notable Examples of Eukaryotic Expression Systems

Several eukaryotic expression systems have been successfully utilized for the production of diverse membrane proteins. Here are a few notable examples that showcase the versatility of these systems:

Yeast (Saccharomyces cerevisiae): Yeast has been a workhorse in membrane protein expression, particularly for eukaryotic transporters and channels. The Baker's yeast, S. cerevisiae, offers a well-established genetic toolkit and robust growth characteristics. Notable examples include the expression of the human aquaporin-1 (hAQP1), which helped elucidate the water transport mechanism across cell membranes.

Insect Cells (Sf9 and Sf21): Baculovirus-mediated expression in insect cells, such as Spodoptera frugiperda-derived Sf9 and Sf21 cells, has been instrumental in producing challenging membrane proteins, including GPCRs. The insect cell system was pivotal in obtaining the crystal structure of the β2-adrenergic receptor, a prototypical GPCR.

Mammalian Cells (HEK293, CHO): Mammalian expression systems, including Human Embryonic Kidney (HEK) 293 and Chinese Hamster Ovary (CHO) cells, are preferred for the production of complex membrane proteins, such as ion channels and GPCRs. The expression of the cystic fibrosis transmembrane conductance regulator (CFTR) in HEK293 cells, for example, led to insights into the protein's structure and function in cystic fibrosis.

Pichia pastoris: This yeast species offers advantages such as strong promoters and the ability to perform eukaryotic PTMs. Pichia has been used for the expression of diverse membrane proteins, including the bacterial outer membrane protein OmpF and the human serotonin transporter (hSERT).

Advances in Eukaryotic Expression Systems

Recent advances in eukaryotic expression systems have further expanded their utility in membrane protein research. These advancements address some of the traditional challenges and enhance the flexibility of these systems:

Synthetic Biology Approaches: Synthetic biology tools allow for the fine-tuning of gene expression in eukaryotic systems. Researchers can now design custom promoters, codon optimization sequences, and regulatory elements to optimize membrane protein expression.

CRISPR-Cas9 Genome Editing: Genome editing techniques like CRISPR-Cas9 enable the generation of cell lines with specific modifications to facilitate membrane protein expression and improve stability. This technology has been particularly valuable in the study of disease-associated mutations in membrane proteins.

Cell-Free Expression Systems: While not a traditional eukaryotic system, cell-free expression systems that utilize eukaryotic cell lysates have gained popularity. These systems offer rapid and cost-effective alternatives for membrane protein expression and can be tailored for specific PTMs and labelling strategies.

Eukaryotic expression systems have emerged as indispensable tools for the production of membrane proteins, enabling researchers to delve deeper into their structural and functional intricacies. Despite the challenges associated with these systems, their ability to mimic the native cellular environment and support proper folding and post-translational modifications makes them essential in the pursuit of understanding the role of membrane proteins in health and disease. With ongoing advancements in synthetic biology and genome editing, the future of eukaryotic expression systems holds even greater promise for unravelling the mysteries of these vital cellular components.

4.4 Challenges and Considerations in Membrane Protein Expression

Membrane proteins, essential components of biological membranes, play a pivotal role in numerous cellular functions, making them prime candidates for structural and functional studies. However, obtaining sufficient quantities of pure and functional membrane proteins for such investigations remains a formidable challenge. This challenge is largely attributed to the hydrophobic nature of membrane proteins and their propensity to aggregate when removed from their natural lipid environment. In this section, we delve into the intricacies of membrane protein expression and explore the multifaceted challenges and considerations researchers encounter in their quest to produce these vital biomolecules.

Hydrophobicity and Aggregation

One of the foremost challenges in membrane protein expression lies in their inherent hydrophobicity. Membrane proteins are replete with hydrophobic amino acids, forming transmembrane domains that anchor them within lipid bilayers. When these proteins are expressed in heterologous systems, removed from their native lipid environment, they are prone to misfolding and aggregation. This hydrophobic mismatch between the protein and its aqueous surroundings can lead to the formation of non-functional aggregates, hindering purification and subsequent structural studies.

To mitigate this issue, researchers employ various strategies. Detergents, such as Triton X-100 or n-dodecyl-β-D-maltoside (DDM), can be used to mimic the lipid bilayer environment and solubilize membrane proteins during extraction. However, the choice of detergent is critical, as different membrane proteins may require specific detergents for stabilization. Moreover, the concentration of detergent must be carefully controlled to prevent protein denaturation or aggregation.

Toxicity and Membrane Disruption

Another significant challenge associated with membrane protein expression is the potential toxicity of certain membrane proteins to host organisms. Some membrane proteins, especially those involved in transport processes, can disrupt the cellular environment by altering ion gradients or membrane integrity. This disruption can lead to cell death or interfere with the expression of other essential proteins.

To address this issue, researchers often employ tightly regulated expression systems that allow for the controlled induction of membrane protein expression. For instance, the use of inducible promoters in bacterial expression systems, such as the T7 promoter, enables precise control over the timing and level of protein production. Additionally, fusion tags or signal sequences can be employed to direct the expressed protein to specific cellular compartments or membrane fractions, reducing its toxic effects.

Post-Translational Modifications

Many membrane proteins undergo post-translational modifications (PTMs) crucial for their functionality. These modifications include glycosylation, phosphorylation, and lipidation. Expressing membrane proteins in heterologous systems may not guarantee the faithful replication of these PTMs, potentially leading to non-functional or misfolded proteins.

To address this challenge, researchers have developed eukaryotic expression systems, such as yeast or mammalian cells, which possess the machinery required for complex PTMs. These systems allow for more accurate recapitulation of native PTMs and are particularly valuable when studying membrane proteins involved in cell signalling, cell adhesion, or immunity. However, they come with their own set of challenges, including increased complexity, cost, and lower protein yields compared to bacterial systems.

Membrane Protein Toxicity and Host Viability

The expression of membrane proteins, especially when overexpressed, can have detrimental effects on host cell viability. This is particularly true for bacterial expression systems, where the overproduction of hydrophobic membrane proteins can lead to the formation of inclusion bodies and cell lysis. Consequently, researchers must strike a delicate balance between achieving high protein yields and maintaining host cell viability.

In addressing this challenge, researchers have developed strategies such as co-expression with molecular chaperones or the use of strains engineered to mitigate membrane protein toxicity. Chaperones, such as GroEL/GroES in E. coli, can assist in proper protein folding, preventing aggregation and inclusion body formation. Additionally, the optimization of growth conditions, including temperature, induction time, and culture medium composition, can significantly impact host cell viability and protein yield.

Protein Stability and Conformational Flexibility

Maintaining the stability and native conformation of membrane proteins during expression and purification is crucial for downstream structural and functional studies. However, membrane proteins often adopt multiple conformations and exhibit conformational flexibility, which can complicate efforts to isolate a single, functional state.

Researchers employ a range of techniques to address this challenge, including the use of stabilizing mutations or truncations, as well as the introduction of disulfide bridges to lock proteins in specific conformations. Additionally, advances in structural biology techniques, such as cryo-electron microscopy (cryo-EM), have allowed for the study of membrane proteins in their native lipid environment, preserving their dynamic behaviour.

Membrane Mimetics and Nanodisc Technology

Recent innovations in membrane protein expression have focused on the use of membrane mimetics and nanodisc technology. These approaches involve reconstituting membrane proteins into nanoscale lipid bilayers, allowing them to be studied in a more native-like environment.

Nanodiscs, in particular, have gained popularity due to their versatility and ability to stabilize a wide range of membrane proteins. By encasing a membrane protein in a nanodisc, researchers can retain its native lipid environment, mitigating many of the challenges associated with detergent-based solubilization. Nanodisc technology has been instrumental in advancing the structural and functional characterization of membrane proteins, especially those with complex lipid interactions.

The expression of membrane proteins is a multifaceted attempt fraught with challenges, ranging from hydrophobicity and aggregation to toxicity and post-translational modifications. Overcoming these obstacles requires a combination of innovative expression systems, strategic protein engineering, and an intimate understanding of the specific membrane protein's biology. As advancements in membrane protein expression continue to evolve, researchers are better equipped to unlock the mysteries of these vital biomolecules, furthering our understanding of cellular processes and contributing to the development of new therapeutic interventions.

Chapter 5: X-ray Crystallography of Membrane Proteins

5.1 Principles of X-ray Crystallography

In the complex world of structural biology, X-ray crystallography stands as a remarkable technique, a beacon illuminating the three-dimensional landscapes of biological molecules, including the elusive membrane proteins. In this section, we embark on a journey through the principles of X-ray crystallography, uncovering the fundamental mechanisms that enable scientists to visualize the intricate structures of these integral components of cellular life.

The Genesis of X-ray Crystallography

The roots of X-ray crystallography trace back to the early 20th century when scientists began to harness the power of X-rays. Wilhelm Röntgen's discovery of X-rays in 1895 marked the inception of this groundbreaking technique. However, it was not until the pioneering work of Sir William Henry Bragg and his son, Sir William Lawrence Bragg, that X-ray crystallography found its wings.

The Bragg duo, in 1912, proposed the Bragg's Law, a fundamental equation in X-ray crystallography. This law elucidated the relationship between the angle of incidence, the wavelength of X-rays, and the distances between the atomic planes in a crystal lattice. Bragg's Law opened the door to deciphering the atomic structures of crystalline materials, including biological macromolecules.

From Crystals to Patterns: The Basics

X-ray crystallography is grounded in the premise that a crystal, a regular and repeating arrangement of atoms or molecules, can diffract X-rays in a predictable manner. The process unfolds as follows:

Crystal Formation: The first step involves growing a crystal of the molecule of interest. In the context of membrane proteins, this is often the most challenging aspect of the entire attempt. Membrane proteins tend to be hydrophobic, making them reluctant to crystallize in an aqueous environment. Researchers employ various techniques, such as lipidic cubic phase crystallization or bicelle methods, to coax these proteins into forming crystals.

X-ray Exposure: Once a crystal is obtained, it is subjected to a beam of X-rays. These X-rays are typically produced using a synchrotron, a highly specialized facility that generates high-intensity X-ray beams. The crystal is placed on a goniometer, allowing precise control of its orientation relative to the X-ray source.

Diffraction Pattern: As the X-rays penetrate the crystal, they interact with the electrons in the atoms within. This interaction results in the scattering of X-rays in different directions. The scattered X-rays form a diffraction pattern on a detector, which is a critical piece of the puzzle in X-ray crystallography.

Cracking the Code: Analysing Diffraction Patterns

The diffraction pattern obtained is akin to a cryptic code, containing valuable information about the arrangement of atoms within the crystal. Deciphering this code involves several key components and concepts:

Spatial Frequencies: The diffraction pattern consists of spots, the intensity and position of which encode information about the crystal's structure. The position of these spots relates to the angles at which X-rays were scattered, while the intensity corresponds to the number of electrons in the atoms at those positions.

Reciprocal Space: A fundamental concept in X-ray crystallography is reciprocal space. The diffraction pattern is converted into reciprocal space using mathematical transformations. In reciprocal space, the spatial frequencies become vectors, simplifying the analysis.

Fourier Transform: Fourier transform is a mathematical operation used to convert the information in reciprocal space back into real space. This operation allows scientists to construct an electron density map, which represents the distribution of electrons in the crystal.

Phase Problem: One of the major hurdles in X-ray crystallography is the phase problem. The diffraction pattern provides amplitude information (intensity) but lacks the crucial phase information, which is essential to reconstruct the electron density map accurately. Several techniques, including multiple isomorphous replacement and anomalous dispersion methods, have been developed to solve this problem.

Building the Molecular Jigsaw: Model Building and Refinement

Once the electron density map is obtained, the next steps involve fitting a model of the molecule into the map and refining it to achieve the best fit. This process can be likened to assembling a complex jigsaw puzzle, where each atom is a piece. Key aspects of this phase include:

Model Building: Researchers use specialized software to manually or automatically build an initial model of the molecule within the electron density map. The model is typically derived from known structures of homologous proteins.

Refinement: The initial model is refined iteratively through computational procedures to improve its fit with the electron density. The refinement process adjusts the positions of atoms and optimizes their interactions to minimize discrepancies between the model and the experimental data.

Validation: Rigorous validation procedures are applied to assess the quality of the final structural model. This includes checking the fit between the model and the electron density map, analysing stereochemistry, and ensuring the absence of steric clashes.

Bridging the Gap: From Crystals to Membrane Proteins

Membrane proteins present unique challenges in X-ray crystallography due to their hydrophobic nature and tendency to aggregate. Researchers have devised innovative strategies to overcome these obstacles. For instance, membrane proteins are often stabilized within a lipidic environment, mimicking their natural habitat. The use of nanodiscs, liposomes, or bicelles helps maintain their structural integrity and facilitates crystallization.

Furthermore, advancements in microfocus X-ray beams and data collection techniques have made it possible to work with smaller and more challenging crystals, including those of membrane proteins. Cryogenic cooling, where crystals are flash-frozen in liquid nitrogen, has become a standard practice to reduce radiation damage and improve data quality.

Challenges and Limitations

While X-ray crystallography has yielded invaluable insights into the structures of membrane proteins, it is not without limitations. Some challenges and considerations include:

Crystallization Difficulty: Membrane proteins can be notoriously challenging to crystallize, and obtaining high-quality crystals can be a laborious process.

Radiation Damage: Exposure to X-rays can cause radiation damage to the crystals, leading to alterations in the crystal lattice and potentially affecting the accuracy of the structure.

Size and Flexibility: Some membrane proteins, especially large or highly flexible ones, may pose additional challenges during data collection and model building.

Non-crystalline Membrane Proteins: Not all membrane proteins can be crystallized. Alternative methods such as cryo-electron microscopy (Cryo-EM) may be more suitable for such cases.

X-ray crystallography is a powerful tool that has revolutionized our understanding of membrane protein structures. It has provided detailed insights into the architecture of these essential cellular components, paving the way for drug discovery and a deeper understanding of their functions. Overcoming the challenges posed by membrane proteins has required creativity and innovation, and the field continues to evolve, pushing the boundaries of what is possible in structural biology.

5.2 Challenges in Crystallizing Membrane Proteins

In the chase of unravelling the intricacies of membrane proteins through X-ray crystallography, researchers confront a myriad of formidable challenges that have, for decades, made this scientific attempt akin to solving a complex puzzle. These challenges, often requiring creative solutions and persistent effort, underscore the unique nature of membrane proteins and their propensity to resist crystallization. In this section, we delve into the diverse hurdles that researchers encounter when attempting to crystallize these vital biomolecules.

Hydrophobic Nature of Membrane Proteins

One of the defining characteristics of membrane proteins is their hydrophobicity. Being embedded within the lipid bilayer, these proteins possess hydrophobic amino acid residues that prefer the company of lipids over water. This inherent hydrophobicity poses a significant challenge during the crystallization process. Unlike their soluble counterparts, membrane proteins resist dissolving in the aqueous solutions typically used for crystallization trials. This reluctance to disengage from the lipid environment often results in aggregation and precipitation, rendering traditional crystallization techniques ineffective.

Lipid-Protein Interactions

In addition to their hydrophobic nature, membrane proteins engage in intricate interactions with surrounding lipids, which further complicate crystallization efforts. These lipid-protein interactions are essential for the stability and functionality of membrane proteins in their native environment. However, they can disrupt crystallization attempts by hindering the formation of well-ordered protein crystals. Lipids may partially cover the hydrophilic surfaces of membrane proteins, preventing solvent molecules and crystallization agents from accessing critical regions for crystal growth.

Detergent Selection and Micelle Formation

To overcome the hydrophobicity challenge, researchers typically employ detergents to solubilize membrane proteins. Detergents serve as amphipathic molecules that can encapsulate hydrophobic regions of membrane proteins, allowing them to exist in a soluble form in aqueous solutions. However, selecting the right detergent is a delicate task. The choice of detergent can significantly impact the stability and conformational integrity of the protein. Moreover, excessive detergent concentrations can disrupt the protein's structure or promote micelle formation, where detergent molecules aggregate around the protein, forming a protective shell that prevents crystallization.

Protein Flexibility and Conformational Heterogeneity

Another challenge in crystallizing membrane proteins stems from their inherent flexibility and conformational heterogeneity. Unlike many soluble proteins that exist in a single well-defined conformation, membrane proteins often adopt multiple conformations to perform their functions efficiently. These dynamic behaviours make it challenging to trap a single, stable protein conformation suitable for crystallography. Researchers must address this challenge by optimizing conditions that stabilize the desired conformation while suppressing conformational heterogeneity.

Sample Purity and Homogeneity

The success of membrane protein crystallization hinges on the purity and homogeneity of the protein sample. Even trace contaminants or minor impurities can disrupt crystallization by introducing heterogeneous nucleation sites or interfering with crystal growth. Achieving high sample purity is especially demanding when working with membrane proteins, as their extraction and purification often involve intricate and time-consuming processes.

Sample Size and Concentration

Crystallization requires a sufficient quantity of protein at the appropriate concentration to generate crystals suitable for X-ray diffraction analysis. However, membrane proteins are often available in limited quantities due to challenges in expression and purification. This limitation poses a critical challenge, as researchers must carefully balance the need for higher protein concentrations with the risk of destabilizing the protein or promoting unwanted aggregation.

Membrane Mimetics and Environment

To facilitate crystallization, scientists often employ membrane mimetics, such as lipidic cubic phases or bicelles, to recreate the lipid bilayer environment in which membrane proteins naturally reside. These mimetics help solubilize the protein and provide a more native-like environment. Nevertheless, the choice of membrane mimetic and its optimization remain significant challenges. The precise conditions required for successful crystallization can vary greatly among different membrane proteins, necessitating extensive screening and optimization efforts.

Choice of Crystallization Method

The choice of crystallization method is pivotal in membrane protein structural biology. Traditional vapor diffusion methods, which work well for soluble proteins, may not be suitable for membrane proteins due to their hydrophobicity. Alternative methods, such as lipidic cubic phase crystallization, lipid monolayer crystallization, or bicelle crystallization, have been developed to address these challenges. Selecting the most appropriate method for a particular membrane protein often involves trial and error, further complicating the process.

Post-Crystallization Challenges

Even after successfully obtaining crystals of a membrane protein, challenges persist. Crystal handling and data collection can be particularly demanding. Membrane protein crystals are often fragile and prone to damage during manipulation, requiring specialized techniques for harvesting and cryoprotection. Additionally, collecting high-quality X-ray diffraction data from these crystals can be challenging due to their inherent anisotropy and radiation sensitivity.

Crystallizing membrane proteins presents a formidable set of challenges, primarily driven by their hydrophobic nature, interactions with lipids, and structural flexibility. Researchers in the field employ a combination of innovative approaches, including the use of detergents, membrane mimetics, and optimized crystallization methods, to overcome these hurdles. Despite the complexities involved, recent advancements in the structural biology of membrane proteins demonstrate that with dedication, ingenuity, and collaborative efforts, researchers continue to make significant strides in deciphering the structures and functions of these vital biomolecules.

5.3 Successful Examples of Membrane Protein Crystallography

In the relentless quest to decipher the three-dimensional structures of membrane proteins, X-ray crystallography has emerged as a paramount tool, enabling us to unravel the secrets hidden within the lipid bilayers. This section delves into notable achievements, showcasing the remarkable successes of membrane protein crystallography.

Rhodopsin - A Pioneer in the Field

One of the groundbreaking milestones in membrane protein crystallography was the elucidation of the structure of rhodopsin, a light-sensitive G-protein coupled receptor (GPCR) found in the rod cells of the retina. This seminal work was conducted by Palczewski et al. in 2000, ushering in a new era of structural biology of membrane proteins. Rhodopsin's structure provided invaluable insights into the seven-transmembrane helical fold, which is a hallmark of GPCRs.

The rhodopsin structure was determined at a resolution of 2.8 Å using X-ray crystallography. The success of this attempt can be attributed to several factors, including advances in protein expression, crystallization techniques, and data collection methods. The breakthrough not only deepened our understanding of visual signal transduction but also paved the way for the structural determination of numerous other GPCRs.

Bacterial Rhodopsin - A Versatile Photoreceptor

Bacterial rhodopsins, such as bacteriorhodopsin and proteorhodopsin, have also played a pivotal role in the development of membrane protein crystallography. Bacteriorhodopsin, discovered in Halobacterium salinarum, is a light-driven proton pump. The determination of bacteriorhodopsin's structure by Dieter Oesterhelt and Walther Stoeckenius in 1971 was the first major success in the structural biology of membrane proteins. This feat was achieved with 7 Å resolution using electron diffraction.

Proteorhodopsin, found in marine bacteria, was another significant milestone. In 2003, three independent research groups simultaneously reported the crystal structures of proteorhodopsin, showcasing the rapid progress in the field. This light-driven proton pump revealed a similar seven-helical transmembrane architecture to bacteriorhodopsin, albeit with variations in the protein's structure.

Potassium Channels - Gatekeepers of Cellular Excitability

Potassium channels are a diverse family of membrane proteins responsible for regulating the flow of potassium ions across cell membranes. Among them, the voltage-gated potassium channels (Kv channels) are vital for controlling neuronal excitability and muscle contraction. In 2003, MacKinnon and colleagues reported the crystal structure of the Kv1.2 potassium channel, a groundbreaking achievement that illuminated the intricate mechanisms governing ion conductance.

This structure, determined at 2.4 Å resolution, revealed the distinctive pore domain and voltage-sensing domains. It demonstrated the selectivity filter's role in ion selectivity and the molecular basis of channel gating. MacKinnon was later awarded the Nobel Prize in Chemistry in 2003 for his contributions to the field, underscoring the significance of this structural revelation.

AQP1 - The Water Channel

Aquaporins, a family of membrane proteins, facilitate the rapid and selective passage of water molecules across cell membranes. Among them, Aquaporin-1 (AQP1) stands as an iconic example. In 1999, Agre and colleagues reported the crystal structure of AQP1, marking a milestone in our understanding of water transport across membranes.

AQP1 forms a tetrameric assembly with each monomer comprising six transmembrane helices and two half-helices. The protein's structure revealed a water-selective pore, formed by a narrow hydrophobic constriction, which prevents the passage of ions and other solutes. This remarkable specificity in water transport has vital implications for cellular osmoregulation, renal function, and numerous physiological processes.

The Potassium Channel Complex - KirBac3.1

KirBac3.1 is a bacterial inward-rectifying potassium channel found in Burkholderia pseudomallei. In 2009, a research team led by Roderick MacKinnon reported the crystal structure of KirBac3.1, offering new insights into ion channel architecture. The structure, determined at a resolution of 3.65 Å, revealed a unique fold that differed from other potassium channels, underscoring the structural diversity within this protein family.

The KirBac3.1 structure provided insights into the mechanisms of ion selectivity and rectification, shedding light on the intricate processes by which cells regulate ion flow across membranes. Additionally, it served as a platform for studying ion channel pharmacology and drug design.

The Powerhouse Complex - Cytochrome c Oxidase

Cytochrome c oxidase (CcO) is a critical enzyme in the electron transport chain, residing in the inner mitochondrial membrane. It plays a central role in oxidative phosphorylation and energy production. In 1996, the groundbreaking crystal structure of bovine heart CcO was reported by Iwata et al., representing a landmark achievement in membrane protein crystallography.

The structure, determined at a resolution of 2.8 Å, showcased the enzyme's catalytic core and its intricate arrangement of prosthetic groups. This revelation deepened our understanding of electron transfer and proton pumping in the mitochondrial respiratory chain, shedding light on the bioenergetics of living cells.

The Remarkable Potassium Pump - Sodium-Potassium Pump (Na+/K+ ATPase)

The sodium-potassium pump, or Na+/K+ ATPase, is an indispensable membrane protein responsible for maintaining cellular ion balance. In 2020, the Nobel Prize in Chemistry was awarded to Benoît Roux, Sir Peter Ratcliffe, and William Kaelin Jr. for their groundbreaking work on the crystal structure of the sodium-potassium pump.

The structure of Na+/K+ ATPase revealed the molecular intricacies underlying ion transport and established a structural framework for understanding the pump's function. This breakthrough holds great promise for advancing our knowledge of ion transport processes and the development of therapeutic interventions for various diseases.

These exemplary instances underscore the remarkable progress made in the field of membrane protein crystallography. From the pioneering days of bacteriorhodopsin to the recent breakthroughs in GPCRs and ion channels, these structures have enriched our understanding of cellular processes and opened avenues for drug discovery and therapeutic interventions.

The success stories presented here represent just a fraction of the multitude of membrane protein structures that have been elucidated using X-ray crystallography. They serve as beacons of inspiration, illuminating the path forward for researchers embarking on the challenging but rewarding journey of deciphering the intricate structures of membrane proteins.

5.4 Recent Advances in X-ray Crystallography

In the dynamic field of structural biology, X-ray crystallography has long held a position of paramount importance. It is a powerful technique, offering an unparalleled lens into the inner workings of biological macromolecules. Over the years, X-ray crystallography has undergone a series of remarkable advancements, making it a cornerstone of structural determination for membrane proteins and other complex biomolecules. In this section, we will explore some of the recent breakthroughs and innovative techniques that have catapulted X-ray crystallography to the forefront of structural biology.

Serial Crystallography

One of the most significant developments in X-ray crystallography in recent years has been the emergence of serial crystallography. Traditionally, crystallographers have relied on single, static crystals for data collection. However, this approach poses significant limitations, especially for challenging membrane protein crystals, which can be fragile and sensitive to radiation damage. Serial crystallography revolutionizes this paradigm by allowing researchers to collect data from a multitude of microcrystals, each exposed to X-ray pulses for a fraction of a second.

This method has opened new vistas for studying membrane proteins. Researchers can now harness microcrystals that were previously deemed unsuitable for traditional crystallography due to size or quality issues. Serial crystallography has been particularly advantageous for studying dynamic processes, such as protein conformational changes or reactions occurring within crystals.

For instance, in a groundbreaking study published in *Nature* in 2018, scientists used serial crystallography to capture snapshots of a membrane protein in action. They investigated the reaction cycle of a light-driven proton pump, bacteriorhodopsin, by collecting data from thousands of microcrystals exposed to X-ray pulses. This allowed them to piece together a molecular movie of the protein's conformational changes during its function, shedding light on the intricate mechanisms underlying its biological role.

Microfocus Beamlines

The advent of microfocus beamlines represents another significant stride in X-ray crystallography. Beamlines are essential facilities where X-ray data are collected from protein crystals. In traditional beamlines, the X-ray beam is relatively large and often too broad for the small and delicate crystals of membrane proteins. However, microfocus beamlines concentrate the X-ray beam to a tiny, focused point, making them ideal for studying microcrystals or crystals with limited size.

These microfocus beamlines have substantially improved the data quality and resolution achievable for membrane protein crystals. Researchers can now extract high-resolution structural information from small and challenging crystals that would have been inconceivable a few years ago. This development has been a game-changer for the field, as it expands the scope of membrane protein studies, particularly when dealing with precious or rare samples.

For example, in a study published in the *Proceedings of the National Academy of Sciences* in 2020, scientists employed a microfocus beamline to determine the crystal structure of a human G protein-coupled receptor (GPCR), a class of membrane proteins involved in signalling and drug targeting. The receptor crystals were minute and prone to radiation damage, but the focused X-ray beam allowed for precise data collection, resulting in a high-resolution structure critical for drug discovery efforts.

Time-Resolved Crystallography

Time-resolved crystallography is a pioneering technique that has reshaped our understanding of biological processes at the atomic level. This approach enables scientists to capture structural snapshots of proteins in action, revealing how they change over time. While it has been employed extensively in soluble protein crystallography, its application to membrane proteins has been challenging due to the aforementioned issues related to crystal size and quality.

Recent innovations in time-resolved crystallography have begun to bridge this gap. By using serial crystallography in conjunction with ultrafast X-ray pulses from free-electron lasers (FELs), researchers can now study membrane proteins in their dynamic states. FELs produce extraordinarily intense X-ray pulses that are thousands of times brighter than synchrotron X-ray sources. This extreme brightness, combined with serial crystallography, allows for the collection of diffraction data from microcrystals before they are damaged by the X-rays.

An illustrative example of the potential of time-resolved crystallography for membrane proteins can be found in a study published in *Science* in 2018. Scientists employed an FEL to investigate the light-activated proton pump proteorhodopsin. By triggering the protein's reaction with a precisely timed laser pulse and collecting diffraction data in femtosecond time intervals, they captured the structural changes occurring during the proton-pumping process. This approach revealed crucial insights into the protein's function and provided a glimpse into the ultrafast dynamics of membrane proteins.

Advanced Data Processing and Software

In parallel with experimental advancements, there have been remarkable strides in the computational aspects of X-ray crystallography. Powerful software tools and algorithms have been developed to tackle various challenges in data processing, structure determination, and refinement. These tools are indispensable for handling the ever-increasing complexity and size of membrane protein structures.

One notable advancement in data processing is the development of Bayesian approaches for crystallographic phasing. Bayesian methods provide a probabilistic framework for solving the phase problem, a critical step in determining the electron density map from X-ray diffraction data. These methods have been particularly beneficial for solving the structures of membrane proteins where obtaining high-quality phases can be challenging.

Additionally, machine learning techniques have gained prominence in crystallography. Deep learning algorithms, in particular, have been applied to predict electron density maps, automate model building, and improve refinement processes. These approaches have significantly accelerated the structure determination process, making it more accessible and efficient.

Integration of Multiple Structural Techniques

In recent years, an exciting trend has emerged: the integration of multiple structural techniques to gain a more comprehensive understanding of membrane proteins. While X-ray crystallography provides high-resolution static structures, other methods, such as cryo-electron microscopy (cryo-EM) and NMR spectroscopy, offer complementary insights into dynamics and membrane protein conformational changes.

Combining these techniques allows researchers to explore membrane protein structures and functions with unprecedented depth. For example, researchers have employed a hybrid approach that combines X-ray crystallography and cryo-EM to study the full-length structure of a membrane protein complex, revealing both its high-resolution static structure and its dynamic flexibility in a lipid membrane environment.

In a recent study published in *Nature Communications*, this integrative approach was used to investigate a challenging multi-subunit membrane protein complex involved in energy production. The researchers first obtained a high-resolution crystal structure of the complex using X-ray crystallography. Then, they used cryo-EM to visualize the complex in a native-like lipid environment, capturing its dynamic conformations during energy transduction. This integrative approach provided a holistic view of the protein's structure and function, highlighting the synergistic power of combining structural techniques.

Final Thoughts

The recent advances in X-ray crystallography have transformed the field of structural biology, especially in the realm of membrane proteins. Serial crystallography, microfocus beamlines, time-resolved crystallography, advanced data processing, and the integration of multiple structural techniques have collectively expanded the horizons of what can be achieved. These innovations have not only allowed researchers to overcome longstanding challenges but have also opened new avenues for investigating the complex workings of membrane proteins.

As X-ray crystallography continues to evolve, it brings us closer to deciphering the intricate molecular mechanisms that govern the functions of membrane proteins. These advancements are not confined to the laboratory but have profound implications for various fields, including drug discovery, bioengineering, and our understanding of fundamental biological processes.

In drug discovery, the ability to determine high-resolution structures of membrane proteins has expedited the design and development of novel pharmaceuticals. Membrane proteins are often targeted by drugs due to their pivotal roles in diseases such as cancer, cardiovascular disorders, and neurodegenerative conditions. The detailed structural information obtained through advanced X-ray crystallography techniques facilitates the rational design of therapeutic agents that precisely interact with specific regions of the target proteins, thereby minimizing side effects and improving drug efficacy.

Moreover, in the realm of bioengineering, these advancements have fuelled innovations in the design of membrane protein-based sensors and nanodevices. By understanding the atomic-level structures and dynamics of these proteins, researchers can engineer them for various applications, such as biosensing, drug delivery, and biofuel production. For instance, the structural insights gained from X-ray crystallography can guide the design of engineered membrane proteins that function as efficient channels for transporting specific molecules or ions, paving the way for tailored biotechnological solutions.

Furthermore, the comprehensive knowledge obtained through advanced X-ray crystallography techniques contributes to our fundamental understanding of life processes. Membrane proteins play critical roles in cellular communication, energy production, and transport of molecules across biological membranes. The ability to visualize these proteins at high resolution and in different functional states enhances our insights into the mechanisms that underlie these essential processes, ultimately deepening our comprehension of life at the molecular level.

The recent advances in X-ray crystallography have transformed the landscape of structural biology, especially concerning membrane proteins. Serial crystallography, microfocus beamlines, time-resolved crystallography, advanced data processing, and the integration of multiple structural techniques have collectively propelled our ability to unveil the intricacies of these vital biomolecules. These advancements have not only surmounted historical challenges but have also catalysed breakthroughs in drug discovery, bioengineering, and our fundamental understanding of biology. As researchers continue to push the boundaries of X-ray crystallography, we can anticipate even more remarkable discoveries on the horizon, promising to unravel the secrets of membrane proteins and the vital roles they play in life's needlepoint.

Chapter 6: NMR Spectroscopy of Membrane Proteins

6.1 Introduction to NMR Spectroscopy

In the exploration of membrane protein structures and dynamics, Nuclear Magnetic Resonance (NMR) spectroscopy stands as a venerable and indispensable technique. Its potency in delivering a nuanced understanding of these biomolecules' atomic-level intricacies has led to its widespread adoption in structural biology. In this section, we embark on a journey to elucidate the fundamental principles that underpin NMR spectroscopy, examining how it brings the inner workings of membrane proteins to light.

The Essence of NMR Spectroscopy

At its core, NMR spectroscopy exploits the magnetic properties of certain atomic nuclei, specifically those possessing a magnetic moment, or "spin." Among the most commonly probed nuclei are hydrogen (1H), carbon (13C), and nitrogen (15N). When placed in a magnetic field, these nuclei align themselves in specific orientations, creating a non-uniform distribution of nuclear spins.

NMR spectroscopy hinges on the fact that when subjected to a radiofrequency (RF) pulse, nuclei transition between spin states. Upon cessation of the RF pulse, nuclei return to their equilibrium positions, emitting radiofrequency signals in the process. These emitted signals, or free induction decays (FIDs), bear vital information about the local chemical environment and spatial relationships of the nuclei.

The Role of Chemical Shifts

One of the foundational concepts in NMR spectroscopy is the chemical shift, denoted as δ. The chemical shift of a nucleus, typically expressed in parts per million (ppm), arises from its local electronic environment. It reflects the nucleus's resonance frequency in comparison to a reference compound, usually tetramethylsilane (TMS) for 1H and 13C nuclei. In the context of membrane proteins, chemical shifts provide a wealth of information about the amino acid residues' environments, helping researchers identify secondary structural elements and conformational changes.

For instance, α-helical segments within membrane proteins typically exhibit distinctive chemical shifts, allowing for the recognition and delineation of these crucial structural motifs. As a result, chemical shift data form the cornerstone for NMR-based structural investigations.

The Power of Resonance Multiplicity

NMR spectroscopy's potency lies not only in the chemical shifts but also in resonance multiplicity, stemming from the interactions between neighbouring nuclei. The most prominent example is the spin-spin coupling, referred to as scalar coupling or J-coupling. When two nuclei are coupled, the resultant NMR spectrum exhibits multiple peaks or splitting patterns.

In the realm of membrane protein studies, spin-spin coupling provides invaluable insights into molecular connectivity. Researchers can deduce the proximity and interaction between specific atoms within the protein sequence, unveiling the intricate web of intra- and intermolecular contacts.

Spatial Information via NOEs

In the elucidation of membrane protein structures, Nuclear Overhauser Effects (NOEs) come to the fore. NOEs arise from dipolar interactions between nuclear spins and can provide crucial spatial information about the proximity of specific nuclei in three-dimensional space.

By analysing NOE data, researchers can discern distances between protons within the protein, facilitating the creation of distance constraints. These constraints, in turn, serve as essential building blocks for computational methods that determine the protein's structure.

Dynamics Unveiled by Relaxation Times

Beyond static structures, NMR spectroscopy excels in capturing dynamic aspects of membrane proteins. This capability arises from the measurement of relaxation times, specifically longitudinal ($T1$) and transverse ($T2$) relaxation times. $T1$ reflects the rate at which excited nuclei return to their equilibrium states, whereas $T2$ relates to the timescale of dephasing of the nuclei's precession.

In the context of membrane proteins, these relaxation times furnish insights into protein dynamics, revealing motions on timescales ranging from picoseconds to milliseconds. Researchers can unravel conformational changes, the flexibility of specific regions, and ligand binding events, all of which are pivotal in understanding protein function.

Experimental Setups in NMR Spectroscopy

Practical NMR experiments demand meticulous sample preparation and precise instrumentation. For membrane proteins, which are often challenging due to their hydrophobic nature, specialized approaches are required. Two-dimensional (2D) and three-dimensional (3D) NMR experiments, where multiple nuclei are correlated, enable the determination of detailed structural information.

Furthermore, magic angle spinning (MAS) NMR has emerged as a powerful technique to study membrane proteins embedded in lipid bilayers. MAS NMR obviates the need for crystallization and allows investigations of proteins in their native environments, akin to their functional conditions within the lipid membrane.

The Impact of Isotopic Labelling

Isotopic labelling plays a pivotal role in enhancing the NMR signal-to-noise ratio and simplifying spectra interpretation. By selectively incorporating stable isotopes (e.g., 15N or 13C) into specific amino acid residues or segments of the protein, researchers can isolate regions of interest and obtain high-resolution spectra.

Isotopic labelling also aids in the collection of heteronuclear data, where correlations between different nuclei (e.g., 1H-15N or 1H-13C) further enrich structural insights. This approach enables the assignment of NMR resonances, crucial for the subsequent determination of the protein's three-dimensional structure.

Limitations and Future Prospects

While NMR spectroscopy offers a wealth of structural and dynamic information for membrane proteins, it is not without its limitations. The technique is sensitive to protein size, and larger proteins often pose challenges in terms of spectral resolution. Additionally, sample conditions, such as protein concentration and stability, can impact data quality.

Looking ahead, technological advancements, such as dynamic nuclear polarization (DNP) NMR and the development of higher-field NMR instruments, hold promise in overcoming some of these limitations. These innovations are expected to expand the horizons of membrane protein research and bring even more of the cellular needlepoint into focus.

NMR spectroscopy is an indispensable tool in the structural biology of membrane proteins, offering a unique window into their atomic-level details and dynamic behaviour. By harnessing the principles of magnetic resonance, researchers can unravel the mysteries of these vital biomolecules, inching closer to a comprehensive understanding of their functions within the cellular milieu.

6.2 NMR Spectroscopy of Membrane Proteins

Nuclear Magnetic Resonance (NMR) spectroscopy stands as an indispensable tool in the arsenal of structural biology. Its application has extended beyond the realm of soluble proteins to encompass the study of membrane proteins, offering a unique window into their structural and functional dynamics. In this section, we will explore how NMR spectroscopy has evolved to tackle the challenges posed by the hydrophobic environment of membranes and elucidate the structure and dynamics of these vital biomolecules.

The Hydrophobic Challenge

Membrane proteins are inherently challenging subjects for structural analysis due to their hydrophobic nature, which limits their solubility in the aqueous environments typically used in NMR experiments. To overcome this obstacle, researchers have developed ingenious strategies.

Detergent Micelles and Bicelles

One approach involves the use of detergent micelles or bicelles to mimic the native membrane environment. Detergents are amphiphilic molecules that can encapsulate membrane proteins, providing a hydrophilic exterior while solubilizing the hydrophobic regions. This approach allows membrane proteins to be studied in a more native-like environment.

For instance, a study on the membrane protein OmpX demonstrated the use of dodecylphosphocholine (DPC) micelles as a suitable medium for NMR analysis. The micelles enveloped the hydrophobic transmembrane region of OmpX, enabling the collection of high-quality NMR data. Similarly, bicelles, composed of lipids and detergents, have been employed to mimic the bilayer environment, preserving the native conformation and function of membrane proteins.

Solubilization with Amphipols

Another approach involves the use of amphipols, amphiphilic polymers that can solubilize membrane proteins while providing a hydrophilic shield. Amphipols have been employed to study a wide range of membrane proteins, including the bacterial protein KcsA and the G-protein-coupled receptor (GPCR) rhodopsin. These studies showcase the versatility of amphipols in preserving the structural integrity of membrane proteins for NMR analysis.

Protein Engineering

In some cases, protein engineering techniques have been employed to enhance the solubility of membrane proteins. This involves the introduction of solubilizing tags or mutations that facilitate the purification and NMR analysis of these proteins. For example, the fusion of a soluble maltose-binding protein (MBP) tag to the target membrane protein has proven effective in enhancing its solubility for NMR studies.

High-Resolution Magic Angle Spinning (HR-MAS) NMR

High-Resolution Magic Angle Spinning (HR-MAS) NMR is another powerful technique used in membrane protein research.

This method involves spinning the sample at the magic angle (54.7 degrees) relative to the magnetic field, reducing line broadening and allowing for high-resolution spectra to be obtained from solid-state NMR experiments. HR-MAS NMR has been applied to membrane proteins incorporated into lipid bilayers or nanodiscs, providing valuable insights into their structure and dynamics.

The Multidimensional NMR Approach

The multidimensional nature of NMR spectroscopy allows researchers to dissect the complex spectra of membrane proteins and extract detailed structural information. Several key experiments are commonly employed in membrane protein NMR studies:

1D 1H NMR Spectroscopy

One-dimensional proton NMR spectra provide valuable information about the chemical shifts and relaxation properties of membrane proteins. While 1D spectra alone may not yield high-resolution structural details, they serve as a starting point for more advanced experiments.

2D Heteronuclear Correlation Spectroscopy

Two-dimensional heteronuclear correlation spectroscopy (2D HSQC) is a fundamental NMR experiment in membrane protein research. It correlates proton and heteronuclear nuclei signals (e.g., 15N or 13C) and provides valuable information about the chemical environment of specific atoms within the protein.

3D and 4D NMR Experiments

To achieve higher resolution and more comprehensive structural information, researchers often employ three-dimensional (3D) and four-dimensional (4D) NMR experiments. These experiments involve additional dimensions, such as carbon (13C) and nitrogen (15N), providing greater spectral dispersion and enabling the determination of long-range contacts within the protein.

Resonance Assignments and NOE Data

Resonance assignments, the process of associating specific NMR signals with individual atoms in the protein, are crucial for structural analysis. Nuclear Overhauser Effect (NOE) data, derived from interatomic nuclear distances, are used to calculate the three-dimensional structure of the membrane protein. Advanced software tools facilitate the automation of these assignments and structure calculations.

Dynamic Nuclear Polarization (DNP) NMR

Dynamic Nuclear Polarization (DNP) NMR is an emerging technique that enhances the sensitivity of NMR experiments. By transferring the high polarization of electrons to nearby nuclei, DNP NMR allows for the detection of weaker signals, enabling the study of larger and more complex membrane proteins.

Paramagnetic Relaxation Enhancement (PRE) NMR

Paramagnetic Relaxation Enhancement (PRE) NMR is a powerful approach for studying the structure and dynamics of membrane proteins in the presence of paramagnetic probes. These probes, when strategically attached to the protein, induce changes in NMR relaxation rates, providing information about local structural fluctuations and protein-lipid interactions.

Solid-State NMR Spectroscopy

Solid-state NMR spectroscopy has emerged as a valuable tool for studying membrane proteins embedded in lipid bilayers or solid supports. It provides structural information about the protein in its native lipid environment, making it particularly relevant for membrane protein research.

Thus, NMR spectroscopy has evolved significantly to meet the challenges posed by membrane proteins. Through the use of innovative solubilization techniques, multidimensional experiments, and advanced technologies like HR-MAS, DNP, and PRE NMR, researchers can now obtain detailed insights into the structure and dynamics of these vital biomolecules. As the field of NMR spectroscopy continues to advance, it promises to unlock further secrets of membrane protein function and pave the way for the development of novel therapeutics targeting this diverse class of proteins.

6.3 Overcoming Challenges in NMR Studies of Membrane Proteins

Nuclear Magnetic Resonance (NMR) spectroscopy stands as a potent technique in the realm of structural biology, offering a unique window into the three-dimensional architecture and dynamics of biomolecules, including membrane proteins. Yet, its application to membrane proteins has been far from straightforward, necessitating innovative solutions and meticulous protocols to address the myriad of challenges posed by the hydrophobic, lipid-embedded nature of these proteins.

Membrane proteins, by their very definition, reside within the lipid bilayer, where the aqueous environment is separated by the hydrophobic core of the membrane. Consequently, their study through NMR spectroscopy presents several distinctive obstacles, each demanding creative approaches to surmount. Here, we delve into these challenges and explore the strategies employed to overcome them, as we embark on an informative journey through the intricacies of NMR studies of membrane proteins.

Detergent Selection and Micelle Formation

One of the initial hurdles in NMR studies of membrane proteins lies in the choice of detergents for solubilization. Detergents act as essential tools to extract membrane proteins from the lipid bilayer, but their presence in the NMR sample can lead to line broadening and spectral congestion. Selecting an appropriate detergent is pivotal in achieving high-quality NMR data.

For instance, dodecylphosphocholine (DPC) and lauryldimethylamine oxide (LDAO) are popular choices for forming micelles around membrane proteins, creating a more amenable environment for NMR. These micelles effectively encapsulate the hydrophobic regions of the protein, offering a "detergent belt" to shield against aggregation and improve spectral resolution.

Protein Size and Complexity

The size and complexity of membrane proteins pose another formidable challenge. Many membrane proteins are inherently large and comprise multiple domains, making them prone to molecular tumbling, which further broadens NMR signals. The intrinsic flexibility of these proteins can also hinder structural determination.

To address this, researchers employ strategies such as segmental labelling and selective deuteration. Segmental labelling involves isotopic labelling of specific protein segments, focusing NMR analysis on individual domains or regions. Selective deuteration, on the other hand, replaces hydrogen atoms with deuterium in specific parts of the protein, reducing the number of signals and simplifying spectra. These techniques help dissect the complexity of membrane proteins, facilitating their structural elucidation.

Sample Stability and Aggregation

Membrane proteins are notorious for their instability and propensity to aggregate, especially when extracted from their native lipid environment. Maintaining sample integrity during NMR experiments is thus a formidable challenge.

One way to mitigate sample instability is by optimizing buffer conditions and pH, often supplemented with stabilizing additives like glycerol or trehalose. Additionally, the use of nanodiscs or lipid bilayer mimetics provides a more native-like environment, reducing the risk of aggregation and enhancing sample stability during NMR experiments.

Spectral Overlap

Spectral overlap is a recurring issue in NMR studies of membrane proteins, particularly when multiple NMR-active nuclei are involved. The similarity in chemical shifts among resonances can hinder accurate peak assignment, making it challenging to extract meaningful structural information.

To circumvent this problem, researchers employ a variety of multidimensional NMR experiments, such as ^{15}N-^{1}H-HSQC (Heteronuclear Single Quantum Coherence), ^{13}C-^{1}H-HSQC, and NOESY (Nuclear Overhauser Effect Spectroscopy), which spread signals across multiple dimensions. These experiments disentangle resonances and provide a more comprehensive view of the protein's structure, aiding in the unambiguous assignment of peaks.

Lipid-Protein Interactions

The interaction between lipids and membrane proteins is a crucial aspect of their function and stability. However, it can complicate NMR studies, as lipid molecules may contribute signals that overlap with those from the protein. This challenge necessitates approaches to distinguish lipid signals from protein signals.

One solution is to selectively label lipid molecules with isotopes like ^{13}C or ^{15}N, enabling the differentiation of lipid and protein peaks in NMR spectra. Additionally, researchers employ paramagnetic probes or lanthanide-binding tags to selectively perturb lipid signals, making them distinct from protein resonances.

Dynamic Heterogeneity

The dynamic nature of membrane proteins, including motions on timescales ranging from picoseconds to milliseconds, poses a challenge in obtaining high resolution NMR data. Fast internal motions can lead to line broadening and reduced signal intensity.

To address dynamic heterogeneity, researchers utilize a range of NMR relaxation techniques, including T1, T2, and heteronuclear NOE (Nuclear Overhauser Effect) measurements. These experiments provide insights into the timescales and amplitudes of protein motions, aiding in the characterization of dynamic properties and their role in protein function.

Advancements in NMR Technology

The field of NMR spectroscopy has witnessed remarkable technological advancements over the years, many of which have been instrumental in overcoming challenges associated with membrane protein studies. High-field NMR instruments, superconducting magnets, and cryogenic probes have significantly improved sensitivity and resolution, allowing researchers to investigate larger and more complex membrane proteins with greater precision.

Moreover, innovative pulse sequences and data acquisition strategies have emerged to enhance the quality of NMR spectra. For example, transverse relaxation-optimized spectroscopy (TROSY) has been instrumental in studying larger membrane protein complexes, as it mitigates relaxation-induced line broadening.

NMR spectroscopy has evolved into a powerful tool for unravelling the structural and dynamic intricacies of membrane proteins. While these proteins pose a unique set of challenges due to their hydrophobic nature and complexity, researchers have developed ingenious strategies to overcome these hurdles. From judicious detergent selection to the use of advanced NMR techniques and technologies, the pursuit of membrane protein structural elucidation continues to advance, offering invaluable insights into their roles in cellular processes and disease mechanisms. In the chapters that follow, we will explore further developments and applications of NMR spectroscopy in the context of membrane proteins, shedding light on the remarkable progress made in this vibrant field of structural biology.

6.4 Case Studies in NMR Structural Determination

In this section, we investigate into the practical application of Nuclear Magnetic Resonance (NMR) spectroscopy for elucidating the structural intricacies of membrane proteins. NMR is a powerful technique that has yielded remarkable insights into the world of biomolecular structures, and its adaptation to membrane protein research has been nothing short of transformative. We will explore a few compelling case studies where NMR played a pivotal role in deciphering the structures of membrane proteins, shedding light on their functional relevance and dynamic behaviour.

Bacteriorhodopsin - A Pioneer in Membrane Protein NMR Studies

Bacteriorhodopsin, a light-driven proton pump found in the cell membrane of Halobacterium salinarum, stands as one of the early successes of NMR in the realm of membrane proteins. This seven-transmembrane helical protein posed significant structural challenges due to its high hydrophobicity and intricate folding. NMR spectroscopy, coupled with solid-state NMR, allowed researchers to obtain a detailed structural model of bacteriorhodopsin. The breakthrough came in 2000 when the three-dimensional structure of bacteriorhodopsin was unveiled. This study revealed not only the overall fold of the protein but also the proton transport pathway and key residues involved in light-driven proton pumping.

Rhomboid Protease - Unravelling a Unique Membrane Protein Family

Rhomboid proteases are integral membrane proteins with a crucial role in intramembrane proteolysis. Understanding their structure and mechanism has been a challenging attempt due to their dynamic nature and limited structural data. NMR spectroscopy played a pivotal role in elucidating the structure of a rhomboid protease from Escherichia coli. Researchers employed solution NMR techniques, including multidimensional heteronuclear NMR experiments, to probe the conformational dynamics of this protein. By combining NMR data with other biophysical techniques, they provided insights into the substrate recognition and catalytic mechanism of rhomboid proteases, opening up new avenues for drug design targeting this unique protein family.

α-Hemolysin - Pore-Forming Toxin as a NMR Model

α-Hemolysin, produced by pathogenic bacteria like Staphylococcus aureus, is a pore-forming toxin that inserts into host cell membranes. NMR spectroscopy has been instrumental in characterising the structure and dynamics of α-hemolysin and its membrane interactions. Researchers used a combination of solution NMR and solid-state NMR techniques to study the transmembrane domain of α-hemolysin. This allowed them to map out the protein's membrane-embedded regions, revealing its structural changes during pore formation. These insights have not only advanced our understanding of bacterial pathogenesis but also have potential applications in the design of antimicrobial agents.

G-Protein Coupled Receptors (GPCRs) - Key Players in Cell Signalling

GPCRs are a diverse family of membrane proteins that mediate cellular responses to a wide range of signals, including hormones and neurotransmitters. They have long been of great interest due to their involvement in numerous physiological processes and as drug targets. NMR spectroscopy has been employed in GPCR research to provide crucial structural insights. One notable example is the solution NMR determination of the human β2-adrenergic receptor, a prototypical GPCR. This landmark study offered the first high-resolution structure of a GPCR in a lipid bilayer environment. By using carefully designed samples and innovative NMR techniques, researchers successfully characterised the receptor's conformational dynamics and ligand interactions, providing a blueprint for future GPCR studies and drug discovery efforts.

Outer Membrane Proteins (OMPs) - NMR Sheds Light on Antibiotic Resistance

Outer membrane proteins (OMPs) are essential components of the outer membrane in Gram-negative bacteria, and they play a crucial role in antibiotic resistance. Understanding the structures of OMPs is vital for developing new antibiotics that can circumvent bacterial defence mechanisms. NMR spectroscopy has contributed significantly to this area of research. A noteworthy example is the elucidation of the structure of OmpX from Escherichia coli, a protein involved in antibiotic resistance. Solid-state NMR experiments allowed researchers to determine the protein's secondary structure and orientation within the lipid bilayer. This information provided insights into how OmpX contributes to antibiotic resistance and inspired the design of novel antibiotics targeting OMPs.

In each of these case studies, NMR spectroscopy played a pivotal role in overcoming the challenges posed by membrane proteins. It allowed researchers to uncover the structural intricacies, dynamic behaviours, and functional relevance of these proteins. These insights not only advance our fundamental understanding of membrane proteins but also have significant implications for drug discovery, antibiotic development, and our ability to combat diseases. As NMR techniques continue to evolve, they promise to illuminate even more aspects of membrane protein biology, offering new avenues for scientific exploration and therapeutic innovation.

Chapter 7: Cryo-Electron Microscopy of Membrane Proteins

7.1 Basics of Cryo-EM

Cryo-Electron Microscopy (Cryo-EM), a groundbreaking technique in structural biology, has emerged as a powerful tool for studying membrane proteins. Unlike traditional methods such as X-ray crystallography and NMR spectroscopy, Cryo-EM allows researchers to visualize biological macromolecules, including membrane proteins, in their near-native state. In this section, we will delve into the fundamental principles and techniques that underpin Cryo-EM, highlighting its significance in membrane protein research.

The Evolution of Electron Microscopy

To comprehend the significance of Cryo-EM, it is essential to trace its evolutionary path within the realm of electron microscopy. The electron microscope, invented in the early 20th century, revolutionized our ability to observe structures at the nanoscale. Initially, this technology suffered from limitations, particularly concerning the preservation of biological samples. Conventional electron microscopy required the use of high-energy electrons, which had destructive effects on delicate biological specimens, leading to structural alterations.

The Cryo-Preservation Revolution

The advent of Cryo-EM can be largely attributed to the development of cryo-preservation techniques. Cryo-preservation involves flash-freezing specimens to extremely low temperatures, typically in liquid nitrogen or liquid ethane. This process effectively immobilizes biomolecules in their native conformations, preventing the formation of damaging ice crystals. As a result, biological samples can be maintained in a near-native state, a crucial aspect of structural biology.

Sample Preparation for Cryo-EM

One of the critical aspects of Cryo-EM is the preparation of samples for imaging. The process begins with the deposition of a thin layer of the sample onto a grid, typically made of materials like copper or gold. However, in the case of membrane proteins, which are often embedded in lipid bilayers, this step can be particularly challenging. Researchers must strike a delicate balance between preserving the lipid environment and ensuring a homogeneous distribution of the protein on the grid.

Imaging Process

Once the sample is prepared, it is transferred to the electron microscope, which operates in a vacuum to prevent electron scattering. Cryo-EM employs low-energy electrons that are less destructive to biological molecules. These electrons are focused onto the sample, and the interactions result in the production of an image. Unlike traditional electron microscopy, Cryo-EM captures a series of two-dimensional images from different orientations, creating a dataset known as a tilt series.

Data Processing and 3D Reconstruction

The true power of Cryo-EM lies in its ability to generate three-dimensional reconstructions of macromolecules. This is achieved through advanced computational techniques. Each image in the tilt series contains information about the specimen from a specific angle. Using mathematical algorithms, these images are aligned and combined to produce a 3D reconstruction, providing a detailed, high-resolution structure of the biological molecule under investigation.

Achieving High-Resolution Structures

Recent advancements in Cryo-EM have substantially improved its resolution capabilities. The resolution of a Cryo-EM structure is often measured in angstroms (Å), with higher values indicating finer details. In recent years, Cryo-EM has achieved resolutions that rival those of X-ray crystallography and NMR spectroscopy, making it an indispensable tool in structural biology. For membrane proteins, which are notoriously challenging to crystallize, Cryo-EM has become a method of choice.

Applications in Membrane Protein Research

Cryo-EM has made significant contributions to the structural elucidation of membrane proteins. For instance, it has shed light on the structure of membrane transporters, G-protein coupled receptors (GPCRs), and ion channels, all of which play crucial roles in cellular function. Membrane proteins often have dynamic conformations and may exist in multiple states, making

Cryo-EM particularly valuable for capturing their structural diversity.

Challenges and Limitations

While Cryo-EM has transformed the field of structural biology, it is not without its challenges and limitations. One significant limitation is the need for expensive equipment and expertise in sample preparation and data processing. Additionally, obtaining high-resolution structures may require extensive computational resources. Moreover, Cryo-EM is not suitable for all types of samples, particularly those that are small or highly flexible.

Future Directions in Cryo-EM

Despite its challenges, Cryo-EM continues to evolve, driven by ongoing technological advancements. Researchers are continually working to improve data acquisition speed, enhance resolution, and develop new image analysis algorithms. These efforts aim to expand the applicability of Cryo-EM and make it accessible to a broader scientific community.

Cryo-EM has become a cornerstone technique in the structural biology of membrane proteins, enabling researchers to obtain high-resolution structures of these vital biomolecules in their native environments. Its evolution from traditional electron microscopy, coupled with cryo-preservation techniques and sophisticated data processing methods, has opened new frontiers in our understanding of membrane protein structure and function. As technology continues to advance, Cryo-EM is poised to play an even more significant role in unravelling the mysteries of membrane proteins and their roles in cellular processes and disease.

7.2 Sample Preparation for Cryo-EM

In the world of structural biology, few techniques have experienced such a meteoric rise as cryo-electron microscopy (cryo-EM). It has evolved into a cornerstone method for unravelling the intricacies of membrane protein structures. The transformative power of cryo-EM lies not only in its capacity to capture high-resolution images of membrane proteins but also in the meticulous process of sample preparation. This section delves into the pivotal role of sample preparation in cryo-EM studies of membrane proteins, exploring the techniques and considerations that make this possible.

The Vitality of Proper Sample Preparation

Before we venture into the techniques, it's essential to appreciate the profound influence of sample preparation on the eventual success of a cryo-EM experiment. Like a skilled artist preparing their canvas, scientists must carefully craft their samples. In cryo-EM, this process is the canvas upon which the exquisite details of membrane proteins are painted.

Sample preparation isn't a mere formality; it is the very foundation upon which cryo-EM rests. Any shortcomings at this early stage can reverberate throughout the experiment, resulting in compromised data quality and resolution. It's akin to building a grand cathedral; if the cornerstone is weak, the entire structure may crumble. In the realm of structural biology, where precision is paramount, sample preparation is the cornerstone.

The Cryo-EM Sample Triad

A cryo-EM sample isn't a solitary entity; it's a complex triad comprised of three integral components: the specimen itself, the supporting grid, and the vitrified ice layer. To understand the intricacies of sample preparation, we must first dissect these components.

The Specimen: At the heart of every cryo-EM experiment lies the specimen - in our case, the membrane protein of interest. For optimal results, the specimen should be purified to the highest degree possible, free from contaminants that might obscure the protein's native conformation. Moreover, it's vital to ensure the protein remains in a biologically relevant state throughout the process.

The Supporting Grid: Think of the supporting grid as the stage upon which our specimen will perform. It is typically composed of materials like copper, gold, or graphene, with minute holes called holes or grids. The choice of grid material can influence the quality of the data. For instance, graphene grids have garnered attention for their ability to reduce background noise and improve image contrast.

The Vitrified Ice Layer: Surrounding the specimen is a thin layer of vitrified ice. Achieving the right thickness and consistency of this ice layer is critical. If too thick, it can obscure the protein; if too thin, it may not provide sufficient support. Vitrified ice, unlike regular ice, has an amorphous structure without ice crystals, preserving the specimen in a near-native state.

The Art of Grid Preparation

The supporting grid, often a humble copper or gold mesh, plays a crucial role in cryo-EM sample preparation. Achieving the perfect grid requires precision and finesse, akin to a master craftsman fashioning a delicate piece of jewellery.

Firstly, the grid material is selected based on the specific requirements of the experiment. Copper grids, for instance, are known for their stability and robustness, making them a popular choice. Gold grids, on the other hand, offer high-contrast imaging due to their inherent electron density.

Next, the grid must be meticulously cleaned to remove any contaminants that could interfere with the sample. Ultrasonic cleaning in organic solvents is a common practice, ensuring a pristine surface for specimen application.

Now, comes the delicate task of specimen application. A droplet containing the purified membrane protein is carefully pipetted onto the grid's surface. The grid is then blotted to remove excess liquid, leaving behind a thin film of solution containing the specimen. Achieving the right thickness of this film is crucial; too thin, and the protein may not be adequately supported, too thick, and the specimen may be obscured by excess material.

Freezing in the Blink of an Eye

The transformation of our specimen into a vitrified state is where the true magic of cryo-EM happens. This step involves the rapid cooling of the grid, suspending the specimen in a glass-like ice layer, preserving its native conformation for imaging.

The process of vitrification is akin to capturing a fleeting moment in time. Rapid cooling is essential, as slow freezing would lead to the formation of ice crystals, which could damage the specimen. The grid, now loaded with the specimen, is plunged into a bath of liquid ethane or propane cooled to incredibly low temperatures using liquid nitrogen or helium. This flash-freezing process transforms the thin film of solution into vitrified ice, suspending the specimen in its native state.

Achieving the ideal vitrification conditions demands precision and control, much like a chef crafting the perfect soufflé. Even minor deviations in temperature or humidity can significantly impact the outcome. This step is where the 'cryo' in cryo-EM truly comes to life, and it's here that the sample preparation process reaches its zenith.

A Multifaceted Challenge

Sample preparation for cryo-EM of membrane proteins is a multifaceted challenge. It requires the fusion of diverse disciplines, from biology to materials science to engineering. Each step in the process demands attention to detail, precision, and a deep understanding of the unique characteristics of membrane proteins.

Moreover, the process isn't without its idiosyncrasies. Membrane proteins can be particularly delicate and prone to denaturation or aggregation during sample preparation. Therefore, scientists must employ various strategies, such as the use of mild detergents or amphipathic polymers, to maintain the protein's integrity.

Sample preparation is the lifeblood of cryo-EM studies of membrane proteins. It is a symphony of precision, combining the careful purification of the specimen, the artistry of grid preparation, and the flash-freezing alchemy of vitrification. The success of any cryo-EM experiment hinges on this foundation, and the quality of data and insights gained are a testament to the skill and dedication of those who master this art. As cryo-EM continues to unlock the mysteries of membrane protein structures, sample preparation remains at the forefront, shaping the future of structural biology.

7.3 Data Acquisition and Processing

In the kingdom of membrane protein structural biology, the advent of cryo-electron microscopy (cryo-EM) has revolutionized the field, offering a powerful lens through which to examine these elusive biomolecules. This section will delve into the crucial aspect of data acquisition and processing in cryo-EM, shedding light on the intricate techniques involved.

Cryo-Electron Microscopy (Cryo-EM) - A Paradigm Shift

Traditionally, structural biologists primarily relied on X-ray crystallography and nuclear magnetic resonance (NMR) spectroscopy to determine the structures of biomolecules. However, these methods often faced limitations, particularly when applied to membrane proteins due to their hydrophobic nature and their resistance to crystallization. Cryo-EM emerged as a transformative technique that circumvented these challenges, enabling the imaging of proteins in their native, hydrated state.

Data Acquisition: Snapshot of the Microscopic World

At the heart of cryo-EM lies the process of data acquisition, which involves capturing a series of two-dimensional (2D) projection images of the specimen from various angles, effectively generating a three-dimensional (3D) reconstruction. The core components of this technique are the electron microscope, the specimen, and the detector.

Electron Microscope: The electron microscope serves as the primary tool for data acquisition. Unlike optical microscopes, which use visible light, electron microscopes utilize a beam of electrons. These high-energy electrons interact with the specimen, providing higher resolution due to their shorter wavelength. Transmission electron microscopy (TEM) is the most commonly used modality in cryo-EM.

Specimen Preparation: A critical aspect of cryo-EM is specimen preparation. To preserve the native state of membrane proteins, samples are rapidly frozen in vitreous ice. This vitrification process ensures that the biomolecules are immobilized without forming ice crystals, which could damage the specimen. Membrane proteins embedded in a thin layer of amorphous ice are then placed on a specialized grid for imaging.

Detector Technology: Modern cryo-EM benefits from advanced detector technology. Direct electron detectors have largely replaced traditional photographic film due to their superior sensitivity and dynamic range. These detectors capture electrons as they pass through the specimen, converting them into digital images.

The Role of Image Processing

Once data acquisition is complete, the real challenge begins: processing the vast amount of raw data into a meaningful 3D structure. Image processing in cryo-EM is a meticulous and iterative process that involves several key steps:

Micrograph Alignment: Because electron micrographs are inherently noisy, it is crucial to align them accurately. This process corrects for minor imperfections in the microscope, ensuring that each image is in the correct position and orientation.

Particle Picking: In many cases, the membrane protein of interest is surrounded by a sea of other molecules. Particle picking is the process of identifying and extracting individual particles from the micrographs. This step is often semi-automated, utilizing algorithms to locate potential particles.

Image Pre-processing: Raw images may suffer from various aberrations, such as drift and beam-induced motion. Pre-processing corrects these issues and improves the quality of the data.

2D Classification: To enhance signal-to-noise ratio, particles are classified into groups based on their similarities. This step helps to identify subsets of particles with a common orientation.

Initial 3D Reconstruction: Using the 2D class averages, an initial 3D reconstruction is calculated. This serves as a starting point for the subsequent refinement steps.

3D Refinement: The refinement process involves iteratively improving the 3D reconstruction. It adjusts parameters to optimize the fit between the calculated 3D model and the experimental images, enhancing the resolution of the final structure.

Resolution Assessment: Determining the resolution of the final 3D reconstruction is crucial for assessing the quality of the model. Resolution measures how closely two points in the structure can be distinguished and is a key indicator of structural accuracy.

Validation: Validation procedures ensure that the reconstructed 3D model accurately represents the native structure of the membrane protein. This includes comparing the model to other experimental data, such as biochemical and biophysical measurements.

Overcoming Challenges in Data Processing

Data processing in cryo-EM is not without its challenges. First and foremost is the sheer volume of data generated. Cryo-EM experiments can produce terabytes of data, requiring substantial computational resources and storage capacity. Moreover, the process can be computationally intensive and time-consuming.

Additionally, the quality of the final 3D reconstruction is highly dependent on the quality of the input data. Factors such as specimen heterogeneity and drift during data collection can pose significant challenges. As a result, researchers often spend considerable effort on optimizing specimen preparation and data collection protocols.

Advancements in data processing software have been instrumental in addressing these challenges. State-of-the-art algorithms and computational tools have made it possible to extract high-resolution structural information from complex datasets.

Case Studies in Cryo-EM Data Processing

To illustrate the power of cryo-EM data acquisition and processing, consider the case of the ATP sensitive potassium (KATP) channel. This integral membrane protein plays a critical role in regulating insulin secretion and is a target for diabetes treatment. Cryo-EM studies have provided valuable insights into the channel's structure and gating mechanism, contributing to our understanding of diabetes-related processes.

Another remarkable example is the structure determination of the HIV-1 envelope glycoprotein (Env) trimer. Env is a challenging target due to its flexibility and extensive glycosylation. Cryo-EM allowed researchers to visualize the trimer's intricate structure at high resolution, aiding in the development of potential vaccines and therapeutics against HIV.

In the world of membrane protein structural biology, data acquisition and processing using cryo-EM have transformed the landscape. This powerful technique has enabled researchers to delve into the inner workings of these complex biomolecules, providing unprecedented insights into their structures and functions. As technology continues to advance, cryo-EM promises to reveal even more secrets of the membrane proteome, driving innovation in drug design and our understanding of cellular processes.

7.4 Recent Breakthroughs in Cryo-EM of Membrane Proteins

In recent years, the field of structural biology has witnessed transformative breakthroughs in the structural determination of membrane proteins through the application of Cryo-Electron Microscopy (Cryo-EM). This powerful technique has revolutionized our understanding of membrane protein architecture, dynamics, and function, offering insights that were once considered elusive. In this section, we delve into some of the remarkable achievements in Cryo-EM of membrane proteins, highlighting their significance and potential implications.

High-Resolution Structures of Membrane Protein Complexes

One of the most notable advancements in Cryo-EM is the attainment of high-resolution structures of membrane protein complexes. Traditionally, structural biologists faced immense challenges in elucidating the three-dimensional architectures of these intricate complexes due to their inherent flexibility and heterogeneity. However, recent breakthroughs in Cryo-EM have enabled researchers to overcome these obstacles.

A remarkable example is the determination of the structure of the mitochondrial ATP synthase complex. This membrane protein complex, crucial for cellular energy production, proved elusive for many years. Traditional methods, such as X-ray crystallography, struggled to capture its native state. However, Cryo-EM provided a breakthrough by allowing scientists to visualize this complex at near-atomic resolution. This achievement unveiled the intricate molecular machinery responsible for ATP synthesis within the mitochondrial inner membrane.

Similarly, the Cryo-EM structure of the bacterial ribosome, a complex comprised of both membrane and soluble components, has provided unprecedented insights into protein synthesis. This breakthrough has profound implications for the development of novel antibiotics targeting bacterial ribosomes, highlighting the direct relevance of Cryo-EM to human health.

Studying Membrane Protein Dynamics

Another remarkable frontier in Cryo-EM is the study of membrane protein dynamics. Understanding how these proteins undergo conformational changes during their biological functions is essential for deciphering their mechanisms of action. Recent developments in single-particle Cryo-EM have enabled researchers to capture snapshots of membrane protein dynamics at various stages of their functional cycle.

One illustrative example is the structural characterization of G-protein coupled receptors (GPCRs), a family of membrane proteins involved in signal transduction. GPCRs undergo conformational changes upon ligand binding, which trigger downstream signalling events. Cryo-EM has allowed scientists to capture the different conformational states of GPCRs, shedding light on their activation mechanisms. Notably, the Cryo-EM structure of the β2-adrenergic receptor in complex with its G-protein revealed the subtle but critical conformational changes associated with GPCR activation.

Moreover, Cryo-EM has played a pivotal role in understanding the dynamics of membrane transporters. These proteins are central to the movement of ions and molecules across biological membranes. Recent breakthroughs in Cryo-EM have facilitated the visualization of transporters in various conformations, elucidating their transport mechanisms. For instance, the Cryo-EM structure of the glucose transporter GLUT1 provided insights into the alternating access mechanism, where the transporter alternates between inward-facing and outward-facing conformations to facilitate glucose transport.

Membrane Protein Assemblies and Interactions

Cryo-EM has also been instrumental in deciphering the architecture of membrane protein assemblies and their interactions with other cellular components. Membrane proteins often function in complexes, and understanding these interactions is crucial for comprehending their roles in cellular processes.

A striking example is the elucidation of the structure of the nuclear pore complex (NPC), a massive assembly of membrane proteins that regulates nucleocytoplasmic transport. Cryo-EM studies revealed the intricate arrangement of NPC components, allowing researchers to unravel its gating mechanism. This breakthrough not only deepened our understanding of nucleocytoplasmic transport but also has implications for cancer research, as NPC dysregulation is associated with certain malignancies.

Furthermore, Cryo-EM has played a pivotal role in understanding the interaction between membrane proteins and lipids. The Cryo-EM structure of the potassium channel KCNQ1 in complex with lipids provided insights into how lipids modulate channel function. Such findings have broad implications for drug development, as targeting lipid-protein interactions may offer novel therapeutic strategies.

Cryo-EM in Drug Discovery

The ability to visualize membrane protein structures at high resolution has significant implications for drug discovery. Cryo-EM has emerged as a powerful tool for structure-based drug design, allowing researchers to target membrane proteins with unprecedented precision.

For instance, the Cryo-EM structure of the adenosine A2A receptor in complex with an antagonist provided insights into the molecular basis of receptor-ligand interactions. This knowledge has facilitated the rational design of drugs targeting this receptor, with potential applications in the treatment of Parkinson's disease.

Moreover, Cryo-EM has been instrumental in the development of therapeutics targeting ion channels. The Cryo-EM structure of the TRPV1 ion channel in complex with capsaicin, the compound responsible for the heat sensation in chili peppers, unveiled the structural basis of channel activation. This breakthrough has paved the way for the design of novel analgesic drugs that specifically target TRPV1.

Future Prospects

As we look to the future, Cryo-EM is poised to continue its transformative impact on membrane protein structural biology. The ongoing development of hardware, software, and sample preparation techniques promises even higher resolutions and increased throughput. This will enable researchers to tackle more challenging membrane protein targets and gain deeper insights into their structures and functions.

Moreover, the integration of Cryo-EM with other structural and functional approaches, such as NMR spectroscopy and molecular dynamics simulations, holds the potential to provide a comprehensive understanding of membrane protein biology. These multidisciplinary approaches will be invaluable in addressing complex biological questions and advancing drug discovery efforts.

Cryo-Electron Microscopy has ushered in a new era in the structural biology of membrane proteins. Recent breakthroughs have not only expanded our knowledge of these vital cellular components but have also opened up exciting avenues for drug discovery and therapeutic development. As technology continues to advance, Cryo-EM is poised to remain at the forefront of membrane protein research, offering a glimpse into the intricate molecular world that governs cellular life.

This section highlights the transformative impact of Cryo-Electron Microscopy in structural biology, showcasing its significance in unravelling the complexities of membrane proteins and its potential applications in drug discovery and beyond.

Chapter 8: Membrane Protein Function and Dynamics

8.1 Understanding Membrane Protein Functions

Membrane proteins constitute a crucial component of the cellular landscape, orchestrating diverse functions essential for life. From cellular communication to energy production, these proteins are the workhorses of biological systems, bridging the gap between the interior and exterior environments of the cell. In this section, we will explore the multifaceted functions of membrane proteins, highlighting their vital roles in various cellular processes.

Transporters: Shuttling Molecules Across Membranes

One of the fundamental functions of membrane proteins is to facilitate the transport of molecules across cellular membranes. These proteins, known as transporters, play a pivotal role in maintaining the homeostasis of ions, nutrients, and metabolites within the cell. An illustrative example of this is the sodium-potassium pump (Na+/K+ pump) found in the plasma membrane of animal cells. This pump actively transports sodium ions (Na+) out of the cell and potassium ions (K+) into the cell, thereby establishing a concentration gradient essential for nerve impulse transmission and muscle contraction.

Transporters are not limited to ions; they also oversee the movement of sugars, amino acids, and other vital molecules. For instance, the glucose transporter GLUT1 ensures the uptake of glucose, a primary energy source, into red blood cells. Dysregulation of transporter function can lead to severe medical conditions. In the case of the cystic fibrosis transmembrane conductance regulator (CFTR), mutations result in defective chloride ion transport, causing the buildup of thick mucus in the airways and leading to the symptoms associated with cystic fibrosis.

Receptors: Sensing the Extracellular Environment

Membrane proteins are also adept at sensing external signals and relaying them to the cell's interior. Receptor proteins are exemplars of this function. They recognize a diverse array of ligands, including hormones, neurotransmitters, and growth factors, initiating intracellular responses upon binding.

The epidermal growth factor receptor (EGFR) is a classic example of a receptor protein. Upon binding of epidermal growth factor (EGF), EGFR undergoes a conformational change, triggering a cascade of intracellular events that culminate in cell division and growth. Aberrant activation of EGFR is implicated in various cancers, making it a target for cancer therapy.

Enzymes: Catalysing Vital Reactions

Membrane proteins are not confined to passive roles; they are active participants in many biochemical reactions. Enzymes embedded within the lipid bilayer or associated with it catalyse reactions that are critical for cellular metabolism and energy production.

The ATP synthase, found in the inner mitochondrial membrane, is a remarkable example of a membrane-bound enzyme. It harnesses the proton gradient across the membrane to synthesize adenosine triphosphate (ATP), the cell's primary energy currency. This process, known as oxidative phosphorylation, powers most cellular activities. Disruptions in ATP synthase function can lead to a host of disorders, including mitochondrial diseases.

Adhesion Proteins: Maintaining Cellular Structure and Communication

Beyond individual cells, membrane proteins also play pivotal roles in cell-cell and cell-extracellular matrix interactions. Adhesion proteins are crucial for maintaining tissue integrity, enabling cell migration, and facilitating cellular communication.

Cadherins, a family of cell adhesion molecules, are instrumental in holding cells together within tissues. Their calcium-dependent interactions ensure the cohesion of epithelial layers in organs like the skin and intestines. Defective cadherin function can lead to conditions such as pemphigus vulgaris, an autoimmune disorder characterized by the loss of skin cell adhesion.

Channels: Regulating Ion Flow

Ion channels, a subclass of membrane proteins, control the passage of ions across membranes in a highly regulated manner. These channels are critical for processes such as nerve signalling, muscle contraction, and maintaining membrane potential.

The voltage-gated sodium channel (NaV) is a prime example of an ion channel. It opens in response to changes in membrane voltage, allowing the rapid influx of sodium ions and initiating the propagation of nerve impulses. Dysfunctional NaV channels are associated with various neurological disorders, including epilepsy.

Structural Proteins: Shaping the Cell

Membrane proteins are also involved in shaping and stabilizing cellular structures. Cytoskeletal proteins like spectrin, which binds to the inner surface of the plasma membrane, provide mechanical support to the cell and help maintain its shape. In red blood cells, spectrin contributes to their characteristic biconcave shape, allowing them to squeeze through narrow capillaries.

Signalling Complexes: Coordinating Cellular Responses

Many membrane proteins participate in intricate signalling networks by forming complexes with other proteins. For example, G-protein coupled receptors (GPCRs) are a family of receptors that interact with G-proteins upon ligand binding. This interaction initiates a cascade of intracellular events, regulating processes ranging from vision to immune response.

The β2-adrenergic receptor, a GPCR, is the target of beta-blocker drugs used to treat hypertension and heart conditions. By modulating the receptor's activity, these drugs can regulate heart rate and blood pressure.

Membrane proteins are the linchpin of cellular functionality. Their diverse roles, from transporting molecules to transmitting signals and catalysing reactions, underpin the very essence of life. Understanding these functions is not only essential for advancing our knowledge of biology but also for developing targeted therapies for diseases that result from membrane protein dysfunction. In the following sections, we will delve deeper into the dynamics of membrane proteins and the techniques used to uncover their intricate workings.

8.2 Role of Structural Biology in Elucidating Function

In the world of membrane proteins, form invariably dictates function. To discern the precise role that these proteins play in the grand scheme of cellular life, researchers turn to structural biology as their compass, allowing them to navigate the intricate molecular landscapes that are membrane proteins. This chapter elucidates the profound influence of structural biology in revealing the functional secrets hidden within the folds and twists of these crucial biological components.

Structural biology's utility in the realm of membrane protein research cannot be overstated. While the physiological functions of many membrane proteins remain elusive, their structural blueprints provide a roadmap, guiding us towards comprehension. This section underscores how structural biology, through various methodologies like X-ray crystallography, nuclear magnetic resonance (NMR) spectroscopy, and cryo-electron microscopy (cryo-EM), shines a beacon of light on the functional mysteries of membrane proteins.

Structural Biology Unveiling Protein Architecture

Understanding membrane protein function hinges on knowing their 3D structures. Structural biology serves as a key player in deciphering the intricate architecture of these proteins, providing snapshots of their conformations at atomic resolutions. One exemplary case is bacteriorhodopsin, a light-driven proton pump found in Halobacterium salinarum. Through X-ray crystallography, researchers unveiled the precise arrangement of its seven transmembrane helices, bound retinal cofactor, and the proton transport pathway. This structural information illuminated the protein's mechanism of light-driven proton pumping, laying the foundation for subsequent functional studies.

Similarly, integral membrane transporters, crucial for nutrient uptake and waste expulsion, have had their mysteries exposed by structural biology. The crystal structure of the bacterial transporter lactose permease LacY depicted its unique "rocker-switch" mechanism, wherein helical motions facilitate sugar translocation. Such structural insights deepened our understanding of active transport processes and paved the way for targeted drug design.

Functional Insights from Dynamic Snapshots

Membrane protein function often hinges on their dynamic behaviour, making static structures only part of the puzzle. NMR spectroscopy, a technique embraced by structural biologists, offers a window into the dynamic aspects of these proteins. For instance, the study of ion channels, which regulate the flow of ions across membranes, benefits greatly from NMR's capacity to capture transient conformations. By employing NMR, researchers discerned how the conformational dynamics of the KcsA potassium channel underlie ion selectivity, a pivotal aspect of its function.

Moreover, NMR has been instrumental in elucidating G-protein coupled receptor (GPCR) function. These receptors transduce extracellular signals into intracellular responses, making them prime drug targets. NMR's ability to detect subtle conformational changes allowed scientists to observe ligand-induced structural alterations in GPCRs, shedding light on the mechanism of signal transduction.

Cryo-EM Revolutionizing Structural Biology

In recent years, cryo-EM has undergone a revolution in structural biology, drastically improving our ability to resolve the structures of membrane proteins. This technique, which involves freezing biological samples and bombarding them with electrons, has overcome the crystallization challenges that stymied X-ray crystallography for many membrane proteins.

The 2017 Nobel Prize in Chemistry recognized the groundbreaking contributions of cryo-EM to structural biology. This technique has elucidated the structures of numerous challenging membrane proteins. For example, the cryo-EM structure of the TRPV1 ion channel, responsible for heat sensation and pain, unveiled the architecture of its ion-conducting pore. This breakthrough highlighted the role of specific amino acid residues in ion selectivity, elucidating how TRPV1 functions as a thermosensor.

Cryo-EM has also been instrumental in revealing the architecture of large macromolecular complexes, such as the bacterial ribosome in complex with membrane proteins. This knowledge has enhanced our understanding of protein synthesis and transport across membranes.

Structural Insights into Functional Diversity

Structural biology has been pivotal in uncovering the remarkable functional diversity within the membrane protein superfamily. For example, the structural elucidation of aquaporins, water channel proteins, highlighted their unique selectivity filters, explaining their unparalleled ability to facilitate water transport while excluding protons and other ions. Such findings have profound implications in understanding osmoregulation and renal physiology.

Additionally, structural studies of rhodopsin, a GPCR responsible for vision, showcased the structural basis of its light-sensing function. The identification of its retinal-binding pocket and the conformational changes upon photon absorption illuminated the molecular events underpinning vision.

The Functional Role of Lipids

Membrane proteins do not operate in isolation; they interact intimately with lipids in the lipid bilayer. Structural biology has unveiled the significant role of lipids in modulating protein function. For example, the crystal structure of the potassium channel K2P1 in complex with phosphatidylinositol 4,5-bisphosphate (PIP2) revealed how lipids stabilize the open state of the channel, influencing its gating.

Furthermore, structural studies of rhomboid proteases, involved in intramembrane proteolysis, highlighted their dependence on lipid environments for proper functioning. Understanding these lipid-protein interactions is crucial for comprehending the regulation of diverse cellular processes.

Structural biology stands as a cornerstone in the quest to uncover the functional mysteries of membrane proteins. From revealing architectural intricacies to capturing dynamic snapshots, and from elucidating the diversity within the membrane protein superfamily to unravelling the pivotal role of lipids, structural biology equips us with the tools to decipher the language of these enigmatic biomolecules. Through structural insights, we bridge the gap between form and function, unlocking the secrets that membrane proteins guard within their molecular confines. The chapters ahead will continue to explore the multifaceted nature of membrane protein function and dynamics, built upon the foundation of structural revelation.

8.3 Dynamics of Membrane Proteins

The structural elucidation of membrane proteins, while fundamental, represents only one facet of understanding these vital biomolecules. Inextricably linked to their static structures is the dynamic nature of membrane proteins. Dynamics encompass the conformational changes, movements, and fluctuations that membrane proteins undergo to carry out their functions. Unveiling these dynamics is paramount, as they provide insights into mechanisms underpinning diverse cellular processes. In this section, we will explore the significance of studying the dynamics of membrane proteins, techniques used for their investigation, and the pivotal role dynamics play in their functionality.

Significance of Membrane Protein Dynamics

To comprehend why membrane protein dynamics are essential, it's imperative to acknowledge that biomolecular function often hinges on motion. Proteins are not rigid, static entities; they are dynamic machines, perpetually adapting to their environment. Membrane proteins, embedded within the fluidic lipid bilayer, are no exception. Their functions, such as ion transport, signal transduction, and molecular recognition, necessitate structural flexibility.

One remarkable example of dynamic membrane proteins is ion channels. These proteins, which facilitate the passage of ions across the membrane, exhibit conformational changes akin to a gate. For instance, in voltage-gated ion channels like the sodium channel, the opening and closing of the channel is orchestrated by voltage changes, a dynamic process essential for propagating nerve impulses.

Moreover, G-protein coupled receptors (GPCRs), a prolific class of membrane proteins, exemplify the significance of dynamics in signal transduction. These receptors, upon ligand binding, undergo conformational changes that initiate intracellular signalling cascades. Without these dynamic structural alterations, the transmission of signals across the membrane would be impeded, disrupting cellular communication.

Membrane protein dynamics also come to the fore in the context of transporters. Transporters, responsible for moving molecules across the membrane, rely on alternating access mechanisms, where dynamic transitions between multiple conformations facilitate substrate transport. Understanding these dynamic events is imperative for elucidating the intricate mechanisms governing substrate translocation.

Techniques for Studying Membrane Protein Dynamics

Exploring membrane protein dynamics presents a formidable challenge, primarily due to their size, heterogeneity, and the intricacies of the lipid bilayer environment. However, a repertoire of techniques has been developed to dissect these dynamics:

Nuclear Magnetic Resonance (NMR) Spectroscopy: NMR spectroscopy is a powerful tool for investigating the dynamics of membrane proteins in solution. It provides information on picosecond to millisecond timescale motions, shedding light on backbone and side-chain dynamics. In particular, ^{15}N and ^{2}H NMR relaxation experiments have been utilised to elucidate timescale-dependent dynamic behaviour.

Molecular Dynamics (MD) Simulations: Molecular dynamics simulations offer a computational framework to study the dynamic behaviour of membrane proteins at atomic resolution. These simulations allow researchers to observe how membrane proteins flex, bend, and interact with their surroundings over a range of timescales, from femtoseconds to microseconds.

Hydrogen-Deuterium Exchange Mass Spectrometry (HDX-MS): HDX-MS is a valuable technique to probe the solvent accessibility and dynamics of membrane proteins. By monitoring the exchange rates of amide hydrogens with deuterium, researchers can map regions of proteins that are undergoing conformational changes.

Electron Paramagnetic Resonance (EPR) Spectroscopy: EPR spectroscopy, especially site-directed spin labelling EPR, provides insights into the mobility and structural dynamics of specific residues within a membrane protein. By attaching spin labels to strategic locations, researchers can monitor changes in the spin label environment and infer protein dynamics.

Fluorescence Spectroscopy: Fluorescence techniques, including Förster Resonance Energy Transfer (FRET) and Fluorescence Correlation Spectroscopy (FCS), have been employed to investigate membrane protein dynamics. These methods can probe changes in distance, orientation, and interactions between fluorescently labelled domains or proteins.

Role of Dynamics in Functionality

Understanding the dynamics of membrane proteins is not merely an academic pursuit; it holds practical implications. The dynamic nature of these proteins is often intricately tied to their functionality and has significant implications in drug design and therapeutic interventions.

Drug Design: Many drugs target membrane proteins, and an understanding of their dynamics is crucial for rational drug design. Dynamic regions or conformational changes involved in ligand binding can serve as potential drug targets. For example, the dynamic nature of the active site in kinases has been exploited for designing specific inhibitors.

Disease Mechanisms: Perturbations in membrane protein dynamics are implicated in various diseases. Alzheimer's disease, for instance, involves the aggregation of the amyloid-beta protein, a process intimately linked to changes in protein dynamics. Investigating these dynamics can provide insights into the pathogenesis of such diseases.

Antibiotic Resistance: Membrane proteins play vital roles in antibiotic transport and resistance. Understanding the dynamics of bacterial efflux pumps, for instance, can aid in combating antibiotic resistance by designing drugs that hinder their function.

Functional Mechanisms: Dynamics also underlie fundamental biological processes. For example, in the bacterial photosynthetic reaction centre, dynamic conformational changes are essential for efficient energy transfer. Unravelling these dynamics can illuminate the mechanisms of energy transduction in photosynthesis.

The dynamics of membrane proteins are an integral aspect of their functionality. These proteins do not exist in a static vacuum; they dance to the rhythm of molecular forces, responding to changes in their environment. Techniques such as NMR, MD simulations, HDX-MS, EPR, and fluorescence spectroscopy have paved the way for a deeper understanding of membrane protein dynamics. As our comprehension of these dynamics grows, so too does our ability to manipulate them for therapeutic purposes and to unravel the mysteries of cellular processes. Membrane protein dynamics represent a dynamic field of research, with discoveries continually reshaping our understanding of the intricate world within the lipid bilayer.

8.4 Techniques for Studying Protein Dynamics

In understanding the intricate functioning of membrane proteins, it becomes crucial to fathom not just their structural aspects but also their dynamic behaviour. The dynamic nature of membrane proteins plays a pivotal role in their function, which often involves conformational changes, interaction with ligands, and responses to cellular signals. The study of protein dynamics has emerged as an indispensable component of structural biology. In this section, we will explore the techniques employed to unravel the dynamic intricacies of membrane proteins.

Nuclear Magnetic Resonance (NMR) Spectroscopy

NMR spectroscopy stands as one of the foremost techniques for studying protein dynamics at the atomic level. Unlike other structural methods, NMR offers a unique advantage by providing information about both the three-dimensional structure and dynamics of proteins in solution. The technique relies on the interaction between magnetic nuclei (commonly ^1H, ^{15}N, and ^{13}C) and magnetic fields, enabling the observation of atomic positions and motions.

Through NMR, researchers can elucidate the timescales of motions, from picoseconds to milliseconds. This ability to probe timescales is pivotal in understanding membrane protein dynamics because it allows us to identify not only global conformational changes but also local fluctuations within the protein structure.

One powerful NMR approach for studying membrane protein dynamics is **Nuclear Overhauser Effect (NOE) spectroscopy**. NOE experiments provide distance restraints between pairs of nuclei in a protein. These distance restraints can be used to deduce the flexibility and dynamics of specific regions within the protein.

For instance, NMR studies on the membrane protein **Bcl-xL**, a key player in apoptosis regulation, revealed that a flexible loop region undergoes dynamic conformational changes upon ligand binding. This dynamic behaviour is essential for its anti-apoptotic function.

Molecular Dynamics Simulations

Computational methods have revolutionised our ability to explore the dynamics of membrane proteins. **Molecular Dynamics (MD) simulations** are a computational approach that allows researchers to track the movements of individual atoms over time. By applying Newton's laws of motion, MD simulations can reveal the dynamic behaviour of proteins with remarkable detail.

In MD simulations, a model of the protein is constructed, and the forces acting on each atom are calculated iteratively over short time steps. By observing how the protein atoms move and interact, researchers can gain insights into various aspects of dynamics, such as side-chain flexibility, domain movements, and ligand binding events.

For instance, MD simulations were instrumental in understanding the dynamic behaviour of the **bacterial potassium channel (KcsA)**. These simulations revealed how the channel undergoes conformational changes to facilitate ion permeation, shedding light on the molecular basis of ion selectivity and gating.

Hydrogen-Deuterium Exchange Mass Spectrometry (HDX-MS)

HDX-MS is a versatile technique for probing the dynamics of proteins in solution. It involves the exchange of hydrogen atoms in a protein with deuterium atoms from the surrounding solvent. The rate at which this exchange occurs depends on the protein's conformation and flexibility.

By measuring the extent of deuterium uptake at various time points, researchers can deduce the regions of a protein that are exposed to solvent and, therefore, more dynamic. HDX-MS can provide insights into the dynamics of both soluble and membrane proteins.

In a notable application, HDX-MS was employed to study the dynamics of the **β2-adrenergic receptor**, a G-protein coupled receptor (GPCR) embedded in the cell membrane. This technique revealed that ligand binding induces conformational changes not only in the ligand-binding pocket but also in distant regions of the receptor, highlighting the allosteric nature of GPCR activation.

Fluorescence Spectroscopy

Fluorescence spectroscopy is a versatile tool for monitoring protein dynamics. It relies on the interaction between a fluorescent probe and the protein of interest. When the probe is excited with light of a specific wavelength, it emits fluorescence, the intensity and wavelength of which can be monitored to glean information about the protein's conformation and dynamics.

A widely used approach in membrane protein research is **Fluorescence Resonance Energy Transfer (FRET)**. FRET involves the use of two fluorophores – a donor and an acceptor – that are attached to different regions of the protein. When the donor is excited, it can transfer its energy to the acceptor if they are in close proximity (typically within 10-100 Å). By measuring the FRET efficiency, researchers can infer conformational changes and protein dynamics.

For example, FRET was employed to investigate the dynamics of the **rhodopsin**, a light-sensitive GPCR found in photoreceptor cells of the retina. This study revealed that rhodopsin undergoes conformational changes upon light activation, providing crucial insights into the molecular events underlying vision.

Electron Paramagnetic Resonance (EPR) Spectroscopy

EPR spectroscopy is a powerful technique for studying the dynamics of proteins, particularly those with unpaired electrons, such as radicals or metal ions. In EPR, a paramagnetic spin label is introduced into the protein, and the interactions between the label's unpaired electrons and an external magnetic field are probed.

EPR can provide information about the local environment, flexibility, and dynamics of specific regions within a protein. By attaching spin labels at different positions, researchers can obtain a detailed picture of how these regions move and interact.

For instance, EPR spectroscopy was used to investigate the dynamics of the transmembrane helices of the **bacterial transporter EmrE**. This study revealed that the helices exhibit dynamic motions crucial for the protein's function, shedding light on the mechanism of multidrug transport.

Understanding the dynamic behaviour of membrane proteins is fundamental to unravelling their functional mechanisms. The techniques discussed in this section, from NMR spectroscopy and molecular dynamics simulations to hydrogen-deuterium exchange mass spectrometry, fluorescence spectroscopy, and electron paramagnetic resonance spectroscopy, offer diverse tools for probing protein dynamics. Combining these techniques allows researchers to piece together a comprehensive picture of how membrane proteins move and adapt in response to their cellular environment. Such insights not only advance our understanding of fundamental biological processes but also hold promise for drug discovery and the development of targeted therapies aimed at modulating membrane protein dynamics. As technology continues to advance, the study of membrane protein dynamics remains a dynamic and evolving field in structural biology.

Chapter 9: Membrane Protein-Lipid Interactions

9.1 Lipid Bilayers in Membrane Protein Function

In the dynamic world of membrane biology, lipid bilayers serve as the foundational framework upon which the performance of membrane proteins hinges. These amphipathic molecules, consisting of hydrophilic heads and hydrophobic tails, self-assemble into a double-layered structure that envelops and segregates the aqueous interior of cells, forming a semi-permeable barrier that demarcates cellular compartments. This lipid bilayer not only provides structural integrity to the cell but also plays an indispensable role in modulating the function of integral membrane proteins that traverse it. The importance of lipid-protein interactions in regulating protein function is increasingly recognised, and this section will illuminate the profound influence of lipid bilayers on membrane protein function, replete with pertinent examples and empirical evidence.

The Lipid Bilayer: A Dynamic Scaffold

The lipid bilayer's primary role is to compartmentalise the cell, isolating various cellular processes and maintaining cellular integrity. Composed of a diverse array of lipid species, it is far from a static entity; instead, it exists in a perpetual state of flux. The fluid mosaic model, proposed by Singer and Nicolson in 1972, offers an apt description of the bilayer's dynamic nature. In this model, proteins float within the lipid sea, contributing to the membrane's mosaic-like appearance. This fluidity is integral to membrane protein function, as it enables lateral diffusion and the dynamic assembly of protein complexes.

Lipid Composition and Membrane Protein Function

The lipid composition of the bilayer is not uniform across all cellular membranes; it varies depending on the organelle's function. This diversity in lipid composition profoundly impacts the activity and stability of resident membrane proteins. For example, the endoplasmic reticulum (ER) membrane, which houses proteins involved in lipid synthesis, is enriched in phosphatidylcholine and phosphatidylethanolamine, lipids that promote proper protein folding and stability. In contrast, the inner mitochondrial membrane, where the electron transport chain resides, predominantly consists of cardiolipin, a lipid critical for the function of respiratory complexes.

Moreover, specific lipids can directly influence the function of individual membrane proteins. The enzyme HMG-CoA reductase, a key player in cholesterol biosynthesis, is a case in point. This membrane protein is regulated by cholesterol levels in the ER membrane, where it resides. When cholesterol levels are high, it binds to the reductase, leading to its degradation and a subsequent reduction in cholesterol synthesis. Such lipid-protein interactions play a pivotal role in cellular homeostasis.

Lipid Rafts: Microdomains of Functional Significance

Lipid rafts are distinct regions within the lipid bilayer characterised by their enrichment in cholesterol and sphingolipids. These microdomains are highly dynamic and serve as platforms for protein sorting and signal transduction. Their significance lies in their ability to segregate specific membrane proteins, facilitating protein-protein interactions critical for cellular processes.

One well-studied example of lipid raft involvement in membrane protein function is the role of rafts in immune cell signalling. The glycosylphosphatidylinositol (GPI)-anchored proteins, which play a pivotal role in immune cell activation, are localised to lipid rafts. This specific localisation enhances their interactions with downstream signalling molecules, orchestrating an effective immune response. Similarly, G-protein coupled receptors (GPCRs), a large family of membrane proteins that mediate cellular responses to hormones and neurotransmitters, are known to associate with lipid rafts. This association can modulate their signalling properties and responsiveness to ligands.

Membrane Protein Conformational Changes Induced by Lipids

Lipid bilayers have the unique ability to induce conformational changes in membrane proteins. These changes can be subtle, affecting the orientation of specific domains, or more dramatic, leading to alterations in protein structure that are essential for function.

The mechanosensitive ion channels, found in various cell types, provide a compelling example of lipid-induced conformational changes. These channels respond to mechanical forces by opening or closing, allowing ions to pass through the membrane. The lipid bilayer's composition, particularly the presence of certain lipids such as phosphatidylglycerol, influences the channels' sensitivity to mechanical stress. In essence, the lipid environment acts as a sensor that modulates the channels' mechanosensitivity.

Another example lies in the function of rhodopsin, a GPCR responsible for visual perception in photoreceptor cells. Rhodopsin undergoes a light-induced conformational change triggered by the absorption of photons by its bound chromophore, retinal. This conformational change leads to the activation of downstream signalling pathways. However, the surrounding lipid environment, particularly the presence of polyunsaturated lipids in the photoreceptor membranes, influences the stability of the active state of rhodopsin. Thus, the lipid bilayer exerts control over the phototransduction process.

Lipid bilayers are far from passive barriers in the cellular milieu; they are active participants that intricately influence membrane protein function. The lipid composition of the bilayer, the presence of lipid rafts, and the ability to induce conformational changes all contribute to the dynamic interplay between lipids and membrane proteins. Understanding these interactions is not only critical for elucidating fundamental cellular processes but also holds promise for drug development, as many diseases involve membrane proteins whose functions are modulated by specific lipid environments. As we delve deeper into the molecular needlepoint of membrane biology, it becomes increasingly evident that lipids are key players in the grand narrative of cellular function.

9.2 Lipid-Protein Interactions

In the intricate mosaic of the cell membrane, lipids and proteins engage in a dynamic dance, forming a complex choreography that underpins essential biological processes. These lipid-protein interactions are not mere coincidences but finely tuned molecular dialogues crucial for membrane protein function. This section elucidates the nature of these interactions, their significance, and the methods employed to study them.

Nature of Lipid-Protein Interactions

The amphipathic nature of lipids, with hydrophobic tails and hydrophilic heads, makes them ideal partners for membrane proteins. These interactions can be classified into several distinct modes:

Lipid Bilayer Embedment: Many integral membrane proteins are firmly embedded within the lipid bilayer, their hydrophobic transmembrane domains aligning with the fatty acyl chains of lipids. This integration is primarily governed by hydrophobic forces and van der Waals interactions. For instance, the alpha helical transmembrane domains of the G-protein coupled receptor rhodopsin are seamlessly nestled within the lipid bilayer, ensuring its stability and functionality.

Lipid-Exposed Hydrophobic Patches: Some membrane proteins exhibit specific lipid-exposed hydrophobic patches that facilitate interactions with nearby lipids. These regions may function as lipid anchors or participate in conformational changes upon lipid binding. The potassium channel KvAP features a hydrophobic paddle domain that interacts with surrounding lipids, modulating channel gating.

Annular Lipids: Membrane proteins are often encircled by a dynamic 'annulus' of lipids. These annular lipids are in constant flux, associating and dissociating with the protein's surface. They play a vital role in stabilizing the protein structure, preventing aggregation, and influencing its function. For example, in the case of bacteriorhodopsin, annular lipids help maintain its structural integrity, allowing it to function as a light-driven proton pump.

Specific Lipid Binding Sites: Some membrane proteins possess well-defined lipid-binding sites, where particular lipid species interact with high affinity. These interactions can be crucial for protein function. Phospholipid scramblases, for instance, contain binding sites for phosphatidylserine, enabling them to facilitate lipid flip-flopping across the bilayer.

Significance of Lipid-Protein Interactions

Lipid-protein interactions are not merely incidental but are integral to the proper functioning of membrane proteins. Their significance can be observed in several biological contexts:

Protein Stability: Lipids act as scaffolds, stabilizing membrane proteins. This stability ensures the correct folding and structural integrity of these proteins, preventing misfolding and aggregation, which can be detrimental to cellular function.

Conformational Changes: Lipid binding can induce conformational changes in membrane proteins, crucial for their function. For example, in the case of ion channels, lipid interactions can trigger channel gating, allowing the selective passage of ions across the membrane.

Signal Transduction: Lipid-protein interactions play a pivotal role in signal transduction pathways. G-protein coupled receptors (GPCRs) exemplify this, where specific lipid interactions are essential for receptor activation and downstream signalling.

Local Environment: Membrane lipids create a unique microenvironment around membrane proteins. This environment can influence the protein's local dynamics, binding kinetics, and even the accessibility of certain substrates or ligands.

Membrane Protein Trafficking: Lipids can modulate the trafficking and sorting of membrane proteins within the cell. Lipid rafts, for example, are specialized microdomains enriched in cholesterol and sphingolipids, and they facilitate the spatial organization of membrane proteins.

Methods to Study Lipid-Protein Interactions

Elucidating the nuances of lipid-protein interactions is a challenging task, but several experimental techniques have been developed to probe these molecular dialogues:

X-ray Crystallography: High-resolution structures of membrane proteins in complex with lipids can reveal the precise nature of lipid-protein interactions. By co-crystallizing membrane proteins with lipids, researchers have gained insights into binding sites and the orientation of lipids around the protein.

NMR Spectroscopy: NMR spectroscopy is invaluable for studying dynamic lipid-protein interactions. Researchers can use NMR to investigate changes in chemical shifts or relaxation rates of both lipids and proteins upon binding, providing information about binding kinetics and thermodynamics.

Cryo-Electron Microscopy (Cryo-EM): Cryo-EM has emerged as a powerful tool to visualize membrane proteins in their native lipid environment. By vitrifying samples in a thin layer of ice, researchers can capture high-resolution structures of membrane protein-lipid complexes.

Mass Spectrometry: Mass spectrometry techniques, such as native mass spectrometry and lipidomics, allow researchers to analyse the composition of lipid-protein complexes. This method provides quantitative data on the stoichiometry and identity of bound lipids.

Molecular Dynamics Simulations: Computational approaches, such as molecular dynamics simulations, enable researchers to simulate lipid-protein interactions at the atomic level. These simulations provide insights into the dynamic nature of these interactions over time.

Biophysical Assays: Various biophysical techniques, including surface plasmon resonance (SPR) and isothermal titration calorimetry (ITC), are used to quantify lipid-protein binding affinities and thermodynamic parameters.

Lipid-protein interactions are fundamental to the structure, stability, and function of membrane proteins. Understanding these interactions at a molecular level is crucial for deciphering the mechanisms of various cellular processes and for the development of therapeutics targeting membrane proteins. The combination of experimental and computational techniques continues to unravel the intricacies of these vital molecular dialogues, shedding light on the dynamic world within the cell membrane.

9.3 Impact of Lipids on Membrane Protein Structure

In the complex, lipids constitute more than just the background scenery; they play pivotal roles as active participants in the theatre of cellular function. Indeed, the structural intricacies and physiological functions of membrane proteins are profoundly influenced by their lipid surroundings. As we explore this dynamic interplay in this section, it becomes evident that lipids are not mere bystanders but critical actors in shaping the conformation and function of membrane proteins.

The Lipid Bilayer: A Dynamic Environment

The lipid bilayer, which serves as the backdrop for membrane proteins, is not a static canvas but a dynamically fluctuating environment. This dynamic nature arises from the inherent fluidity of lipids. Phospholipids, the primary constituents of the bilayer, have hydrophobic tails and hydrophilic heads, which spontaneously self-assemble into a bilayer in an aqueous environment. The fluidity of this bilayer arises from the ability of lipid molecules to move laterally within the plane of the membrane. The lipid tails undergo rapid rotational and translational motions, akin to a molecular ballet, which imparts fluidity to the membrane.

Lipid Composition: Tailoring the Membrane

One of the key ways in which lipids influence membrane protein structure is through their composition. The diversity of lipid species within a biological membrane is staggering, and this diversity is not arbitrary. Different lipids have distinct chemical properties, including the length and saturation of their acyl chains and the presence of various head groups. These variations can profoundly impact the local microenvironment experienced by membrane proteins.

For instance, the length and saturation of lipid acyl chains influence the thickness and fluidity of the bilayer. Shorter and more unsaturated acyl chains result in a thinner and more fluid membrane, while longer and saturated chains create a thicker and less fluid bilayer. This variation in bilayer thickness can affect the membrane protein's conformation and function.

Membrane proteins may respond to changes in bilayer thickness by adopting different structural states, thereby modulating their activity.

Moreover, the head group chemistry of lipids can influence the electrostatic interactions between the lipids and membrane proteins. Negatively charged head groups, such as those found in phosphatidylserine or phosphatidylglycerol, can interact with positively charged regions on the surface of membrane proteins. These electrostatic interactions can stabilize specific protein conformations or even trigger conformational changes that are essential for protein function.

Lipid-Protein Interactions: Orchestrating Structural Changes

Beyond their passive influence through bilayer properties, lipids actively interact with membrane proteins, orchestrating structural changes. These interactions can be categorized into several types, each with its distinct consequences.

Lipid Annular Shell

In the immediate vicinity of membrane proteins lies a lipid annular shell, a region where lipids surround the protein in a tightly packed arrangement. This annular lipid shell plays a crucial role in stabilizing the protein's structure. For instance, lipids in the annular shell can interact with hydrophobic regions of the protein, preventing water molecules from infiltrating these regions. This hydrophobic shielding helps maintain the protein's stability and prevent unfolding or aggregation.

Lipid-Induced Conformational Changes

Lipids can also induce conformational changes in membrane proteins. An exemplary case is the interaction between G-protein coupled receptors (GPCRs) and their ligands. When a ligand, such as a hormone or neurotransmitter, binds to a GPCR, it can trigger a cascade of events that involve changes in the conformation of the receptor. These conformational changes are often coupled with interactions between the receptor and specific lipid molecules within the membrane. Such lipid-induced conformational changes are central to the function of GPCRs, which are involved in diverse physiological processes, including signal transduction.

Lipid Rafts: Clustering Membrane Proteins

Lipids can also influence the spatial organization of membrane proteins within the bilayer. One notable feature of lipid bilayers is the formation of microdomains known as lipid rafts. Lipid rafts are enriched in specific lipid species, such as sphingolipids and cholesterol, and they have distinct physical properties compared to the surrounding bilayer. These rafts act as platforms for the clustering of membrane proteins.

The partitioning of membrane proteins into lipid rafts can have significant consequences for their function. For example, the clustering of certain receptors within lipid rafts can enhance signal transduction by facilitating the assembly of signalling complexes. Moreover, the lipid composition of rafts can influence the activity of enzymes involved in lipid metabolism, which can, in turn, impact the function of membrane proteins.

Lipid Dynamics and Protein Function

The dynamic nature of lipids in the bilayer can also influence membrane protein function. Lipid molecules in the bilayer undergo rapid lateral diffusion, which means that they can move within the membrane and interact with membrane proteins. This dynamic interaction can impact the function of channels and transporters.

For instance, ion channels, which are essential for the regulation of cellular ion concentrations, can be modulated by the presence of specific lipid molecules. Lipids can bind to ion channels and alter their open and closed states, thereby regulating ion flux. Similarly, transporters that facilitate the movement of molecules across the membrane can be influenced by lipid-protein interactions, which can modulate their transport kinetics.

Lipidomics: Profiling Lipid-Protein Interactions

Advancements in lipidomics, the comprehensive study of lipid species and their interactions, have provided valuable insights into the impact of lipids on membrane protein structure and function. Mass spectrometry-based lipidomics approaches enable researchers to identify and quantify the lipid species present in membranes and to investigate changes in lipid composition under different physiological conditions.

For example, lipidomics studies have revealed that the lipid composition of cellular membranes can be dynamically regulated in response to environmental cues or cellular signalling pathways. Changes in lipid composition can, in turn, influence the behaviour of membrane proteins. This emerging field continues to expand our understanding of how lipids contribute to the structural and functional diversity of membrane proteins.

The impact of lipids on membrane protein structure is a multifaceted phenomenon that encompasses lipid bilayer properties, lipid composition, and specific lipid-protein interactions. Lipids are not passive spectators in the membrane but active participants that sculpt the structural landscape of membrane proteins and modulate their functions. Understanding the intricate interplay between lipids and membrane proteins is essential for deciphering the complexities of cellular processes and holds promise for future therapeutic interventions.

9.4 Lipidomics and Membrane Protein Research

Understanding the intricate interplay between membrane proteins and lipids is paramount in elucidating the functional and structural nuances of these biological entities. Lipidomics, a field that has emerged in recent years, plays a pivotal role in this attempt. It provides a comprehensive and systematic approach to studying the lipids present in biological membranes, shedding light on their diversity, distribution, and dynamics. In this section, we will explore the significance of lipidomics in membrane protein research, examining how it has revolutionised our understanding of membrane biology.

Diversity of Lipids in Membranes

Lipids are not merely passive components of the lipid bilayer; they actively influence membrane protein function and structure. Understanding the lipid composition of membranes is crucial as different lipids exhibit distinct physical and chemical properties. Lipidomics allows us to characterise the lipidome of specific cellular membranes, revealing a staggering diversity of lipid species. For instance, in the inner mitochondrial membrane, lipidomics has unveiled the prevalence of cardiolipin, a unique lipid that plays a pivotal role in the function of respiratory chain complexes, highlighting the specificity of lipid-protein interactions within subcellular compartments.

Dynamic Nature of Lipid-Protein Interactions

One of the remarkable features of lipids in membrane biology is their dynamic nature. Lipids can diffuse within the membrane, affecting the local environment and the function of embedded membrane proteins. Lipidomics, combined with techniques like mass spectrometry, enables researchers to study lipid dynamics with high precision. For instance, in neuronal synapses, the lipid composition changes dynamically during neurotransmitter release and endocytosis, illustrating how lipidomics can help elucidate the temporal aspects of lipid-protein interactions critical for synaptic function.

Mapping Lipid Binding Sites on Membrane Proteins

A key question in structural biology is the precise location of lipid-binding sites on membrane proteins. Lipidomics aids in mapping these binding sites by identifying lipids associated with specific proteins. For instance, in the case of G-protein coupled receptors (GPCRs), a class of membrane proteins with immense pharmacological importance, lipidomics has identified cholesterol as a critical modulator of receptor function. Understanding such interactions can have profound implications for drug design and the development of therapeutic interventions.

Unveiling Lipid-Induced Conformational Changes

Membrane proteins often undergo conformational changes in response to lipid interactions. Lipidomics provides insights into lipid-induced conformational changes, allowing researchers to decipher the mechanisms by which lipids influence protein function. For example, in the case of ion channels, the presence of specific lipids can induce conformational changes that regulate ion flux across the membrane. Lipidomics has contributed significantly to our understanding of the structural dynamics of these channels.

Lipidomics in Disease Research

The role of lipid-protein interactions in disease has garnered considerable attention. Dysregulation of lipid metabolism and lipid-protein interactions is implicated in a wide range of diseases, including cancer, neurodegenerative disorders, and cardiovascular diseases. Lipidomics offers a valuable tool for studying these disease mechanisms. For instance, lipidomics studies have identified alterations in lipid profiles associated with Alzheimer's disease, providing potential biomarkers for early diagnosis and therapeutic targets.

Emerging Techniques in Lipidomics

The field of lipidomics is constantly evolving, with emerging techniques enhancing our ability to probe lipid-protein interactions. Shotgun lipidomics, for instance, enables the rapid and comprehensive profiling of lipids in biological samples. High-resolution mass spectrometry, coupled with advanced separation techniques, has also revolutionised lipidomics by allowing for the identification and quantification of lipid species with unprecedented accuracy. These techniques are instrumental in deciphering the lipid landscapes of membrane systems with increasing complexity.

Challenges in Lipidomics

While lipidomics holds immense promise, it is not without challenges. One major hurdle is the identification of lipid species with high specificity. Many lipids share similar mass spectrometry profiles, making it crucial to develop techniques that can distinguish between them. Additionally, lipid extraction and sample preparation methods can introduce biases, necessitating rigorous standardisation procedures. Moreover, the interpretation of lipidomics data is a complex task, requiring sophisticated bioinformatics tools to unravel the intricate relationships between lipids and membrane proteins.

Future Directions in Lipidomics and Membrane Protein Research

The future of lipidomics in membrane protein research is promising. Advanced lipidomics techniques will continue to provide insights into the roles of lipids in membrane biology, with applications spanning from basic science to drug development. Integration with structural biology techniques, such as cryo-electron microscopy and X-ray crystallography, will enable a more comprehensive understanding of lipid-protein interactions at the atomic level. Moreover, the development of computational models will aid in predicting lipid-protein interactions, facilitating the rational design of membrane-targeted drugs.

Lipidomics has emerged as a cornerstone in membrane protein research, offering a holistic view of the lipid components that orchestrate the structure and function of membranes. It provides the means to decipher the dynamic and intricate interactions between lipids and membrane proteins, paving the way for advancements in both fundamental science and therapeutic development. As technology continues to advance, the synergy between lipidomics and structural biology promises to unravel the mysteries of the lipid-protein interplay in the complex world of biological membranes.

Chapter 10: Drug Targeting and Design of Membrane Proteins

10.1 Membrane Proteins as Drug Targets

In the expansive landscape of drug discovery and development, membrane proteins stand as pivotal protagonists. They constitute an impressive repertoire of pharmaceutical targets, engaged in orchestrating a myriad of cellular processes. While the attempt to decipher their role and structure has been arduous, the tantalizing promise of therapeutic interventions against a multitude of diseases has driven relentless research in this domain.

The Ubiquitous Presence of Membrane Proteins

Before we delve into the therapeutic potential of membrane proteins, it is vital to acknowledge their ubiquity. Membrane proteins are enigmatic molecules that span the cellular membranes, ensuring the segregation and integrity of cellular components. They comprise receptors, channels, transporters, and enzymes, pivotal for cellular communication, nutrient uptake, and myriad other functions.

Integral to cell signalling, G-protein coupled receptors (GPCRs) represent an emblematic example. These receptors, often nestled within the plasma membrane, respond to external stimuli such as hormones and neurotransmitters. Through intracellular signalling cascades, they mediate diverse physiological processes, making them prime targets for drug development. For instance, the β-adrenergic receptor blockers, used to treat hypertension and heart conditions, target GPCRs with remarkable efficacy.

Disease Relevance and Drug Discovery

The perturbation of membrane protein function is frequently linked to pathological conditions, underpinning their significance as therapeutic targets. Consider the case of ion channels, which play a fundamental role in the nervous and cardiovascular systems. Dysfunction in ion channels can lead to disorders such as epilepsy and cardiac arrhythmias. The drug development process targeting ion channels has culminated in the creation of drugs like sodium channel blockers used in epilepsy management.

Furthermore, transporters, such as P-glycoprotein, are integral in drug efflux from cells, influencing drug bioavailability. Their overexpression in certain cancer cells confers multi-drug resistance, posing a formidable challenge in chemotherapy. Research into circumventing P-glycoprotein's efflux activity has the potential to revolutionize cancer treatment.

Structural Insights for Drug Design

One of the key breakthroughs in membrane protein-targeted drug development has been the elucidation of their three dimensional structures. With advancements in structural biology techniques like X-ray crystallography, cryo-electron microscopy, and nuclear magnetic resonance (NMR) spectroscopy, researchers have unveiled the intricacies of membrane protein architecture. These structural insights are akin to discovering a treasure map for drug designers.

A prime example is the structural elucidation of the HIV-1 protease, a viral membrane protein pivotal in the replication of the human immunodeficiency virus (HIV). The structural information enabled the design of protease inhibitors, such as ritonavir and lopinavir, which have transformed HIV from a once-lethal infection to a manageable chronic disease.

Additionally, the structural determination of GPCRs, such as the β2-adrenergic receptor, has led to the rational design of drugs targeting these receptors. For instance, the β2-adrenergic receptor antagonist, salmeterol, is used to treat asthma and chronic obstructive pulmonary disease (COPD). These drugs leverage the knowledge of the receptor's structure to specifically modulate its function, with fewer side effects compared to non-specific drugs.

Challenges and Triumphs in Drug Development

While membrane proteins offer tantalizing prospects, their exploitation as drug targets is rife with challenges. The very attributes that make them appealing, such as their hydrophobic nature and integration within the cell membrane, also pose obstacles. The extraction, stabilization, and crystallization of membrane proteins for structural determination are intricate processes.

Furthermore, designing drugs that selectively target membrane proteins without affecting other cellular components demands precision. Non-specific drug interactions can lead to adverse effects. This challenge was exemplified in the development of antiarrhythmic drugs targeting cardiac ion channels. Many early drugs exhibited proarrhythmic effects, underscoring the importance of structural insights in drug design to avoid unintended consequences.

Despite these hurdles, several remarkable triumphs underscore the potential of membrane proteins as drug targets. For instance, the groundbreaking development of imatinib, a tyrosine kinase inhibitor, revolutionized the treatment of chronic myeloid leukaemia. Imatinib specifically targets the BCR-ABL fusion protein, which drives the proliferation of cancerous cells. Its success exemplifies the power of rational drug design informed by structural knowledge.

Biologics and Membrane Proteins

In the realm of biopharmaceuticals, membrane proteins have also found their niche. Monoclonal antibodies, a class of biologics, have emerged as therapeutic agents against membrane protein targets. Notably, the anti-HER2 antibody, trastuzumab, is employed in breast cancer therapy. HER2, a membrane-bound receptor tyrosine kinase, is overexpressed in certain breast cancers. Trastuzumab specifically binds to HER2, impeding its signalling and inhibiting cancer progression.

The Future of Membrane Protein-Targeted Therapies

The future of drug development undoubtedly holds promise for membrane protein targets. Advancements in structural biology techniques continue to provide high-resolution insights, enabling the design of more selective and potent drugs. Moreover, computational approaches, such as molecular docking and dynamics simulations, complement experimental efforts in drug discovery, expediting the process.

Personalized medicine is another frontier. Tailoring drug therapies to an individual's genetic and membrane protein profile holds potential for enhancing drug efficacy while minimizing side effects. Genomic sequencing and the advent of precision medicine are driving this paradigm shift.

Membrane proteins, with their pervasive role in cellular function and disease, represent a rich needlepoint of drug targets.

Structural insights, gleaned through cutting-edge techniques, have paved the way for rational drug design and the development of biologics. Challenges persist, but the successes in drug discovery against membrane protein targets underscore their significance in shaping the future of pharmacotherapy. As we continue to unravel the intricacies of these remarkable molecules, the pursuit of novel therapeutic interventions remains fervent.

10.2 Rational Drug Design for Membrane Proteins

The rational design of drugs targeting membrane proteins is a captivating area of research at the intersection of structural biology, pharmacology, and medicinal chemistry. Historically, drug discovery often relied on serendipitous findings or high-throughput screening of compound libraries. However, the rational approach takes a more precise and informed route towards drug development. In this section, we explore the principles, strategies, and successful examples of rational drug design for membrane proteins.

Principles of Rational Drug Design

Rational drug design hinges on a deep understanding of the target membrane protein's structure and function. The overarching principle is to design molecules that interact specifically with the protein, modulating its activity in a desired manner while minimizing off-target effects. Here, we discuss the fundamental steps involved in this process:

Target Identification and Validation: The journey begins with the identification and validation of a membrane protein as a potential drug target. This step involves a rigorous assessment of the protein's role in disease, its expression in relevant tissues, and the availability of structural information. Membrane proteins implicated in various diseases, such as G-protein coupled receptors (GPCRs), ion channels, and transporters, have been successful targets for rational drug design.

Structural Characterization: To design drugs rationally, a high-resolution structure of the target membrane protein is indispensable. Techniques like X-ray crystallography, cryo-electron microscopy, and NMR spectroscopy play pivotal roles in providing atomic-level insights into protein structure. For instance, the determination of the crystal structure of the β2-adrenergic receptor paved the way for the development of a new generation of asthma medications.

Binding Site Identification: Once the structure is available, the next step is the identification of binding sites and key residues within the protein. Computational tools, such as molecular docking and molecular dynamics simulations, are invaluable in predicting potential binding sites and estimating binding affinities.

Ligand Design and Screening: With insights into the binding site, medicinal chemists can design molecules (ligands) that are complementary to the site's shape, charge, and hydrophobicity. This process often involves virtual screening of compound libraries to identify potential drug candidates that fit the binding pocket.

Optimization and Testing: Iterative cycles of ligand design, synthesis, and testing are carried out to improve binding affinity, selectivity, and pharmacokinetic properties. Structural information helps in fine-tuning ligand design to maximize interactions with the target protein.

Successful Examples of Rational Drug Design

Rational drug design has yielded remarkable successes in the development of drugs targeting membrane proteins. Here are two notable examples:

Gleevec (Imatinib) - Targeting Tyrosine Kinases: Gleevec, a groundbreaking cancer drug, illustrates the power of rational drug design. It was designed to target a specific fusion protein, BCR-ABL, responsible for chronic myeloid leukaemia (CML). Through X-ray crystallography, scientists obtained the structure of the active site of BCR-ABL, revealing a unique pocket. Imatinib was then crafted to precisely fit this pocket, inhibiting the kinase's activity. Gleevec's design exemplifies the importance of structural insights in developing highly selective drugs.

Tamoxifen - Modulating Oestrogen Receptors: Tamoxifen, used to treat breast cancer, is another triumph of rational drug design. Oestrogen receptors, integral membrane proteins, play a crucial role in breast cancer development. By comprehending the receptor's structure, researchers developed tamoxifen as a selective oestrogen receptor modulator (SERM). Tamoxifen's structure was designed to compete with oestrogen for binding, thereby inhibiting oestrogen's proliferative effects. This rational approach led to a highly effective treatment with fewer side effects.

Challenges in Rational Drug Design for Membrane Proteins

While rational drug design offers exciting prospects, it is not without its challenges, particularly when targeting membrane proteins:

Lack of High-Resolution Structures: One of the primary challenges is the scarcity of high-resolution structures for membrane proteins. Despite advancements in structural biology techniques, some membrane proteins remain elusive, hindering drug design efforts.

Conformational Flexibility: Membrane proteins can exhibit significant conformational flexibility, adopting multiple functional states. Designing ligands that interact with different conformations is a complex task that requires comprehensive structural information.

Lipid Environment: Membrane proteins operate in a lipid bilayer, and the lipid environment can influence their structure and function. Understanding these interactions and their impact on drug binding is a multifaceted aspect of rational drug design.

Transporter Substrate Specificity: Membrane transporters often exhibit strict substrate specificity, making it challenging to design drugs that selectively inhibit or activate them without affecting closely related transporters.

Rational drug design for membrane proteins is a compelling field that harnesses the synergy of structural biology, computational chemistry, and pharmacology. The successes of Gleevec and tamoxifen underscore the potential of this approach in creating highly specific and effective drugs. As technology advances and our understanding of membrane protein structures deepens, we can anticipate a growing number of targeted therapies that offer improved treatment options for a wide range of diseases. The rational design of drugs for membrane proteins is not only an intellectual pursuit but also a transformative force in the realm of drug discovery, poised to benefit countless patients worldwide.

10.3 High-Throughput Screening for Membrane Protein Ligands

In the pursuit of developing therapeutics targeting membrane proteins, researchers face a significant challenge – identifying small molecules or ligands that can modulate the activity of these intricate biological macromolecules. High-throughput screening (HTS) has emerged as a powerful and indispensable tool in this attempt. HTS enables the rapid testing of thousands to millions of compounds for their potential to interact with membrane proteins, ultimately facilitating drug discovery. In this section, we will explore the principles, techniques, and the impact of high-throughput screening on membrane protein research and drug development.

Principles of High-Throughput Screening

High-throughput screening operates on the fundamental premise of testing a vast library of compounds against a target of interest. For membrane proteins, this target might be a receptor, ion channel, or transporter that plays a pivotal role in cellular function or disease. The goal is to identify ligands or compounds that bind to the membrane protein and modulate its activity, either enhancing or inhibiting its function.

One of the key advantages of HTS is its ability to rapidly process a large number of compounds in a relatively short period. This efficiency is essential in the context of membrane protein research, where identifying suitable ligands can be a time-consuming and resource-intensive task. HTS allows researchers to cast a wide net, exploring a vast chemical space for potential hits that can be further optimized into drug candidates.

Techniques in High-Throughput Screening

Several techniques are commonly employed in high-throughput screening for membrane protein ligands. These techniques are designed to detect interactions between the membrane protein and test compounds, typically in a 96- or 384-well plate format. Here are some of the key techniques:

Fluorescence-Based Assays: In this approach, the membrane protein or a suitable label is tagged with a fluorescent marker. When a ligand binds to the protein, it induces a change in fluorescence, allowing for the detection of binding events. For example, researchers have used this method to screen for ligands targeting G-protein coupled receptors (GPCRs), a prominent class of membrane proteins.

Radioligand Binding Assays: Radioactive ligands or tracers are used to detect binding to the membrane protein. When a test compound competes with the radioligand for binding sites, it reduces the amount of bound radioligand, which can be quantified using scintillation counting. This technique has been pivotal in identifying ligands for various ion channels.

Electrophysiological Assays: For ion channels, electrophysiological techniques such as patch-clamp recordings are employed. High-throughput patch-clamp platforms can simultaneously measure the effects of compounds on multiple ion channels, providing valuable data on ion channel modulators.

Label-Free Techniques: Some HTS approaches dispense with the need for labels or tracers. Label-free techniques, like surface plasmon resonance (SPR) or bio-layer interferometry (BLI), directly measure changes in the refractive index or mass as ligands bind to the membrane protein immobilized on a sensor surface.

Impact on Membrane Protein Research and Drug Development

High-throughput screening has revolutionized membrane protein research and drug development in several ways:

Accelerated Drug Discovery: HTS expedites the process of identifying lead compounds for membrane protein targets. The ability to screen thousands of compounds in a short time frame significantly accelerates drug discovery efforts.

Target Validation: HTS can be employed in the early stages of drug development to validate the membrane protein target's relevance in a disease context. Confirming the target's involvement through screening strengthens the rationale for pursuing drug development efforts.

Diverse Chemical Libraries: HTS allows researchers to screen diverse chemical libraries, including natural product extracts and synthetic compounds. This diversity increases the chances of identifying novel, structurally distinct ligands with therapeutic potential.

Structure-Activity Relationship (SAR) Studies: Hits identified through HTS can serve as starting points for structure-activity relationship (SAR) studies. Medicinal chemists can optimize the chemical structure of these hits to improve their affinity, selectivity, and pharmacological properties.

Phenotypic Screening: In addition to target-based screening, HTS can also be used in phenotypic screening. This approach involves screening compounds for their effects on cellular or organismal phenotypes, which can reveal novel membrane protein targets and pathways relevant to disease.

Challenges and Considerations in High-Throughput Screening

While HTS has revolutionized membrane protein research, it comes with its own set of challenges and considerations:

Assay Development: Designing a robust and reproducible assay for a specific membrane protein target can be a complex and time-consuming process. Researchers must consider factors such as protein stability, assay sensitivity, and potential interference from test compounds.

False Positives and Negatives: HTS can generate false-positive and false-negative results. False positives may arise due to non-specific binding, while false negatives can occur if the assay conditions do not suit the target or ligand properties.

Compound Library Selection: The success of HTS depends on the quality and diversity of the compound library. Researchers must carefully curate libraries to ensure they contain compounds relevant to the target and disease of interest.

Hit Validation: Hits identified in an initial HTS campaign must undergo rigorous validation to confirm their binding affinity and biological activity. This often involves follow-up studies using alternative techniques.

Cost and Resources: Implementing HTS requires substantial resources, including specialized equipment, personnel, and compound libraries. Cost-effectiveness and resource allocation must be carefully considered.

High-throughput screening has emerged as a cornerstone of membrane protein research and drug development. Its capacity to rapidly identify potential ligands for membrane proteins has catalysed the discovery of new therapeutics and expanded our understanding of membrane protein function. As technologies continue to evolve, HTS promises to play an even more pivotal role in the development of drugs targeting membrane proteins, addressing a diverse array of diseases and improving patient outcomes.

10.4 Case Studies in Membrane Protein Drug Development

In the complex landscape of drug development, membrane proteins have emerged as promising targets for the pharmaceutical industry. Their intricate roles in cell signalling, transport, and regulation make them ideal candidates for therapeutic intervention. In this section, we will explore several case studies that illustrate the successful development of drugs targeting membrane proteins, shedding light on the pivotal role structural biology has played in these achievements.

Gleevec (Imatinib Mesylate) - Target: BCR-ABL Tyrosine Kinase

One of the most notable success stories in membrane protein drug development is Gleevec, also known as Imatinib Mesylate. Gleevec revolutionized the treatment of chronic myeloid leukaemia (CML) by specifically targeting the BCR-ABL fusion protein, a constitutively active tyrosine kinase responsible for the disease.

Target Protein: The BCR-ABL fusion protein is a membrane-associated tyrosine kinase resulting from a chromosomal translocation. Understanding its structure was essential for drug development.

Structural Insights: Researchers determined the crystal structure of the tyrosine kinase domain of BCR-ABL in complex with Imatinib. This structural information guided the rational design of the drug, enabling it to fit snugly into the kinase's active site, inhibiting its activity.

Clinical Impact: Gleevec achieved remarkable success, turning CML from a life-threatening condition into a manageable chronic disease. It exemplifies the power of structure-based drug design and precision medicine.

Tamiflu (Oseltamivir) - Target: Influenza Virus Neuraminidase

Tamiflu, an antiviral drug used to treat influenza, targets the influenza virus's neuraminidase protein, a membrane glycoprotein essential for viral replication and spread.

Target Protein: Neuraminidase is a membrane protein embedded in the viral envelope.

Structural Insights: X-ray crystallography provided a high-resolution structure of the neuraminidase active site. This information was crucial for designing Oseltamivir, which mimics the natural substrate and inhibits the enzyme.

Clinical Impact: Tamiflu has been a cornerstone in managing influenza outbreaks and pandemics, reducing the duration and severity of symptoms when administered early.

Herceptin (Trastuzumab) - Target: HER2 Receptor

Herceptin, a monoclonal antibody, has transformed the treatment of HER2-positive breast cancer by targeting the human epidermal growth factor receptor 2 (HER2) protein.

Target Protein: HER2 is a transmembrane receptor that plays a key role in cell growth and division. Its overexpression is associated with aggressive breast cancers.

Structural Insights: While Herceptin's development did not rely on a crystal structure of HER2, it exemplifies the potential of monoclonal antibodies to target membrane proteins selectively. Herceptin binds to the extracellular domain of HER2, preventing downstream signalling.

Clinical Impact: Herceptin has significantly improved the prognosis for HER2-positive breast cancer patients, illustrating the efficacy of antibody-based therapies against membrane protein targets.

Ibrutinib (Imbruvica) - Target: BTK Kinase

Ibrutinib is a groundbreaking drug used in the treatment of B-cell malignancies, such as chronic lymphocytic leukaemia (CLL) and mantle cell lymphoma, by targeting Bruton's tyrosine kinase (BTK).

Target Protein: BTK is a cytoplasmic kinase associated with the B-cell receptor signalling pathway.

Structural Insights: Structural biology studies revealed the three-dimensional structure of BTK, aiding in the design of

Ibrutinib. The drug covalently binds to BTK, inhibiting its activity.

Clinical Impact: Ibrutinib has shown remarkable efficacy, offering a targeted treatment option for patients with B-cell malignancies who previously had limited therapeutic choices.

HIV Protease Inhibitors - Target: HIV Protease

The development of HIV protease inhibitors has been a major milestone in the treatment of HIV/AIDS. These drugs target the viral protease, an essential enzyme for HIV replication.

Target Protein: HIV protease is an enzyme located within the viral particle.

Structural Insights: Structural studies using X-ray crystallography provided detailed information about the active site of HIV protease. This knowledge guided the design of protease inhibitors like Saquinavir and Darunavir, which effectively block viral replication.

Clinical Impact: HIV protease inhibitors, when used in combination with other antiretroviral drugs, have dramatically improved the prognosis and quality of life for HIV/AIDS patients.

These case studies underscore the critical role of structural biology in drug development, particularly when targeting membrane proteins. The ability to visualize the three-dimensional structure of these proteins at atomic resolution has empowered scientists to design drugs with remarkable precision, minimizing off-target effects and maximizing therapeutic efficacy. As we move forward, the integration of structural biology techniques will continue to drive advancements in drug discovery and the treatment of diseases influenced by membrane protein dysfunction.

Chapter 11: Membrane Protein Folding and Assembly

11.1 Protein Folding in the Membrane

The process of protein folding within the cellular membrane is a remarkable feat of molecular biology. It underpins the functional integrity of integral membrane proteins, a class of proteins that span the lipid bilayer and play crucial roles in cellular physiology. Understanding the principles governing membrane protein folding is pivotal for unravelling their structures and deciphering their functions. In this section, we will explore the intricacies of protein folding within the hydrophobic environment of the membrane, examining the factors that influence the folding process and the strategies employed by cells to ensure correct folding and assembly.

The Hydrophobic Challenge

Folding a protein within the confines of a lipid bilayer presents a unique challenge due to the inherent hydrophobic nature of the membrane's interior. Unlike the aqueous environment found within the cell's cytoplasm, the membrane's lipid core repels water, creating a hostile environment for most proteins. Integral membrane proteins must adopt a folded conformation that not only fulfils their functional role but also accommodates the hydrophobicity of the membrane.

To adapt to this challenging environment, membrane proteins have evolved a diverse array of structural motifs.

Transmembrane helices, rich in hydrophobic amino acids, are a common feature. These helices span the lipid bilayer, anchoring the protein within the membrane while exposing hydrophilic residues to the surrounding aqueous compartments. This balance between hydrophobic and hydrophilic regions is a key aspect of membrane protein structure and folding.

Co-translational Folding

The process of membrane protein folding begins even before the nascent polypeptide chain emerges from the ribosome. Co-translational folding, as it is termed, involves the simultaneous synthesis of the protein and its insertion into the membrane. The ribosome-associated translocon complex assists in this process, guiding the growing polypeptide chain into the lipid bilayer as it emerges from the ribosome. This co-translational insertion is vital to prevent aggregation and misfolding of hydrophobic segments within the cytoplasm.

An elegant example of co-translational folding is the Signal Recognition Particle (SRP) pathway. Here, a signal sequence at the protein's N-terminus directs it to the ribosome-translocon complex. As the nascent chain elongates, the SRP recognizes the signal sequence and temporarily halts translation. The ribosome-bound protein-SRP complex is then targeted to the endoplasmic reticulum (ER) membrane, where translation resumes. This process ensures that hydrophobic regions are inserted into the membrane as they are synthesized, reducing the risk of misfolding.

Chaperones in Membrane Protein Folding

Just as in the cytoplasm, molecular chaperones play a critical role in guiding membrane protein folding. Chaperones are specialized proteins that assist in the correct folding of their client proteins. In the context of membrane proteins, chaperones are particularly important because they help prevent aggregation and promote the insertion of hydrophobic segments into the lipid bilayer.

One well-studied example of a chaperone involved in membrane protein folding is the SecB chaperone in bacteria. SecB binds to the hydrophobic regions of nascent membrane proteins, preventing premature folding in the cytoplasm. It maintains these proteins in a partially unfolded state, allowing them to be efficiently translocated to the membrane by the SecYEG translocon. Once at the membrane, the protein can then achieve its final folded state.

Role of Post-translational Modifications

Post-translational modifications (PTMs) also influence membrane protein folding. PTMs involve the addition or removal of chemical groups to a protein after translation. In the context of membrane proteins, glycosylation is a common PTM. Glycosylation involves the addition of sugar molecules to specific amino acid residues, often located on the extracellular side of the protein.

Glycosylation can impact membrane protein folding by providing stability and preventing premature folding during biosynthesis. It can also influence protein-protein interactions and play a role in the correct assembly of multi-subunit membrane complexes. An example is the glycosylation of N-linked glycans in glycoproteins, which aids in the correct folding and trafficking of certain membrane proteins.

Protein folding within the membrane is a complex and tightly regulated process that ensures the proper function of integral membrane proteins. The hydrophobic environment of the lipid bilayer necessitates unique strategies, such as co-translational folding and the involvement of chaperones, to guide the folding process. Post-translational modifications, such as glycosylation, further contribute to the stability and functionality of these proteins.

Understanding the nuances of membrane protein folding not only sheds light on fundamental cellular processes but also has practical implications, particularly in the context of drug development. Targeting membrane proteins involved in diseases often requires a deep understanding of their folding pathways and stability. Thus, ongoing research in this field continues to unveil the intricacies of membrane protein folding, offering exciting prospects for both basic science and clinical applications.

11.2 Membrane Protein Topology Prediction

The structural determination of membrane proteins is a complex yet crucial attempt in the field of structural biology. Before embarking on experiments and studies to elucidate the 3D structures of these proteins, it is often necessary to predict their topology within the lipid bilayer. This section explores the methodologies and tools employed for membrane protein topology prediction, offering insights into the techniques that guide researchers in understanding the arrangement of transmembrane domains, orientation, and structural features.

Topology Matters: Understanding Membrane Protein Arrangement

The topology of a membrane protein refers to its specific orientation within the lipid bilayer. It is essential to ascertain the arrangement of transmembrane domains, the location of extracellular and cytoplasmic regions, and the positioning of functional sites. Accurate topology prediction serves as a crucial initial step in membrane protein structural biology, guiding subsequent experiments and aiding in the interpretation of experimental data.

Hydrophobicity Scales and Sequence Analysis

One of the earliest methods employed for membrane protein topology prediction is the analysis of hydrophobicity scales. These scales assign hydrophobic and hydrophilic values to amino acids based on their chemical properties. The premise is that hydrophobic amino acids tend to cluster within the transmembrane regions, while hydrophilic amino acids are more prevalent in the aqueous environment. The Kyte-Doolittle hydrophobicity scale, for instance, has been widely used for this purpose. By scanning the protein sequence for stretches of hydrophobic amino acids, researchers can make initial predictions regarding transmembrane segments.

Furthermore, various computational algorithms have been developed to automate the prediction process. Tools such as TMHMM (TransMembrane prediction using Hidden Markov Models) and HMMTOP employ statistical models and Hidden Markov Models (HMMs) to analyse protein sequences. They consider the statistical likelihood of each amino acid being part of a transmembrane region and predict the number and location of transmembrane segments. These methods have been successful in many cases, providing reasonably accurate predictions. However, they are not without their limitations, particularly when dealing with proteins that exhibit complex topologies or have short transmembrane segments.

Predicting Topology through Experimental Data

While sequence-based prediction methods are valuable, experimental data can offer more direct insights into membrane protein topology. Various techniques, such as site-directed mutagenesis, cysteine-scanning mutagenesis, and accessibility assays, have been used to probe the topological arrangement of membrane proteins.

Site-directed mutagenesis involves the systematic replacement of amino acids with cysteine residues, which can then be chemically modified to probe the environment surrounding them. If a cysteine substitution is accessible from the extracellular side but not the cytoplasmic side, it suggests that the modified residue resides in an extracellular loop, thus contributing to the topology determination.

Cysteine-scanning mutagenesis extends this concept by systematically introducing cysteine substitutions along the entire length of a transmembrane domain. By assessing the reactivity of these cysteine residues to specific chemical reagents, researchers can map the exposure of different regions of the protein to the lipid bilayer or aqueous environment, providing valuable information about the transmembrane topology.

Accessibility assays, such as the protease protection assay, involve subjecting membrane proteins to proteolytic enzymes. Regions of the protein that are protected from digestion by the lipid bilayer will provide insights into the transmembrane segments, while unprotected regions are indicative of exposed domains. These experiments complement the data obtained from mutagenesis studies and enhance our understanding of membrane protein topology.

Structural Data and Topology Prediction

In recent years, the growing availability of high-resolution structural data for membrane proteins has greatly enhanced topology prediction. X-ray crystallography and cryo-electron microscopy provide detailed structural information about the arrangement of atoms within a protein. With this structural data, researchers can unambiguously determine the orientation of transmembrane domains, the location of loops, and the positioning of functional sites.

By superimposing the experimental structures onto the protein sequence, it is possible to derive a precise and experimentally validated membrane protein topology. This approach has revolutionised topology prediction and significantly improved the accuracy of structural models.

Challenges and Future Prospects

While considerable progress has been made in the field of membrane protein topology prediction, challenges persist. The accuracy of sequence-based methods can vary depending on the complexity of the protein, and they may struggle with short transmembrane segments or proteins with unusual topologies. Experimental techniques, while valuable, can be time-consuming and labour-intensive.

Looking ahead, there is ongoing research into the development of more advanced computational methods that incorporate diverse data sources, such as evolutionary information and structural data, to improve prediction accuracy. These approaches aim to provide a more comprehensive understanding of membrane protein topology, even for the most challenging cases.

Predicting the topology of membrane proteins is a fundamental step in understanding their structural biology. Hydrophobicity scales, sequence analysis, experimental mutagenesis, and structural data all contribute to our knowledge in this area. Advances in computational techniques and the increasing availability of high-resolution structural data continue to drive improvements in membrane protein topology prediction, offering researchers a clearer view of these vital cellular components.

11.3 Membrane Protein Oligomerization

The proper functioning of membrane proteins often relies on their ability to interact and form complexes, a process known as oligomerization. These protein-protein interactions play a pivotal role in various cellular processes, from signal transduction to membrane transport. In this section, we will explore the significance of membrane protein oligomerization, the various types of oligomeric assemblies, and examples illustrating their importance in biology.

The Significance of Oligomerization

Oligomerization, the assembly of multiple protein subunits into a functional complex, is a fundamental aspect of membrane protein biology. It provides a structural scaffold for the stability and functionality of various protein families. Membrane proteins often exist in the lipid bilayer as monomers, but they can readily associate with other protein subunits to form oligomers. The oligomeric state of a membrane protein can profoundly impact its function, regulation, and cellular localization.

Types of Oligomeric Assemblies

Membrane protein oligomers come in different flavours, depending on the number of subunits involved and their organization. Here are three common types:

Homodimerization: Homodimerization involves the association of two identical protein subunits. One classic example of homodimerization in membrane proteins is the sodium-potassium pump (Na^+/K^+ pump), which is crucial for maintaining the electrochemical gradients of sodium and potassium ions across the cell membrane. This pump consists of two identical α subunits that form a functional dimer. Each α subunit undergoes conformational changes during ion transport, and the coordinated action of the two subunits is essential for its function.

Heterodimerization: In heterodimerization, two different protein subunits come together to form a complex. An illustrative example is the G-protein-coupled receptor (GPCR) family. GPCRs, which mediate a wide range of cellular responses to external signals, often function as heterodimers. For instance, the GPCR complex formed by the adenosine A1 receptor and the A2A receptor plays a pivotal role in neuroprotection and neurodegeneration. Heterodimerization enables the fine-tuning of signal transduction pathways and receptor pharmacology.

Higher-Order Oligomers: Some membrane proteins form higher-order oligomers, involving more than two subunits. These assemblies can have varying quaternary structures, such as tetramers, pentamers, or larger complexes. One well-known example is the voltage-gated potassium (Kv) channel, a tetramer composed of four α subunits. Kv channels regulate the flow of potassium ions, thereby influencing membrane potential and cellular excitability. The interaction between these subunits is essential for the channel's proper functioning.

Dynamic Nature of Oligomerization

Oligomerization is not a static process; rather, it is dynamic and can be influenced by various factors, including environmental conditions and ligand binding. This dynamism is critical for the adaptability of membrane proteins in response to changing cellular requirements.

For example, the epidermal growth factor receptor (EGFR), a transmembrane receptor with intrinsic kinase activity, can form homodimers upon ligand binding. This dimerization event initiates a signalling cascade, ultimately leading to cell proliferation and differentiation. The dynamic nature of EGFR dimerization allows for tight regulation of cellular responses to growth factors.

Implications for Drug Design and Therapeutics

Understanding membrane protein oligomerization has significant implications for drug design and therapeutics. Many drugs target membrane proteins, and their effectiveness often depends on the oligomeric state of the target protein.

For instance, monoclonal antibodies have been developed to treat cancer by inhibiting the oligomerization of human epidermal growth factor receptor 2 (HER2), a member of the EGFR family. HER2 overexpression is associated with aggressive breast cancer, and monoclonal antibodies like trastuzumab (Herceptin) disrupt HER2 homodimerization, thereby inhibiting cell growth and survival. This exemplifies how targeting oligomerization processes can lead to effective cancer therapies.

Membrane Protein Oligomerization in Cellular Signalling

Oligomerization plays a central role in various cellular signalling pathways. One of the most well-documented examples is the GPCR family. These receptors, upon ligand binding, often form heterodimers or higher-order oligomers to initiate downstream signalling events. This complexity allows for the fine-tuning of signal transduction, as different combinations of receptor subunits can lead to diverse cellular responses.

A prominent example is the opioid receptor system, which includes μ-opioid receptors (MOR), δ-opioid receptors (DOR), and κ-opioid receptors (KOR). These receptors are known to form heterodimers, and the resulting combinations of MOR-DOR and MOR-KOR have distinct functional properties. The oligomerization of opioid receptors modulates the analgesic effects of opioid drugs and contributes to the development of more selective and effective pain management therapies.

Structural Insights into Oligomerization

The elucidation of the structural basis of membrane protein oligomerization has been a major focus of structural biology. Techniques such as X-ray crystallography, NMR spectroscopy, and cryo-electron microscopy have provided invaluable insights into the organization of protein subunits within oligomeric complexes.

For example, the crystal structure of the β2-adrenergic receptor, a GPCR, in complex with its G protein revealed the arrangement of the receptor and G protein subunits within the heterodimeric complex. This structure shed light on the conformational changes that occur upon receptor activation and how they are transmitted to the G protein.

In the case of higher-order oligomers, such as tetrameric ion channels, the structural determination of the Kv channel has provided critical insights into how the four subunits work in concert to control ion flow. The structural studies of such complexes help us understand the mechanistic details of ion channel function and gating.

Membrane protein oligomerization is a fundamental process that underlies the proper functioning of a wide array of proteins involved in vital cellular processes. Understanding the significance, dynamic nature, and structural details of oligomerization is crucial for advancing our knowledge of membrane protein biology and for developing novel therapeutic interventions. As structural biology techniques continue to evolve, we can anticipate even deeper insights into the mechanisms and functional consequences of membrane protein oligomerization. This knowledge will undoubtedly shape the future of drug development and our understanding of cellular signalling pathways.

In the following chapter sections, we will explore the role of chaperones and folding assistance in guiding membrane protein assembly and the exciting field of membrane protein engineering.

11.4 Chaperones and Folding Assistance

In the complex world of membrane protein biology, the journey from the synthesis of a nascent polypeptide chain to its three-dimensional functional form is fraught with perils. The cellular environment is an unpredictable arena, and the emergence of membrane proteins, delicately poised within lipid bilayers, requires a careful orchestration of folding events. In this section, we will explore the critical role of chaperones and folding assistance in guiding these proteins to their correct conformation.

Chaperones: The Guardians of Protein Folding

Imagine a bustling molecular factory where thousands of polypeptide chains are birthed, all destined to play specific roles within the cell. In this frenetic environment, it's easy for a newly synthesized membrane protein to get lost or adopt an incorrect conformation, rendering it dysfunctional or potentially toxic to the cell. This is where molecular chaperones come to the rescue.

Molecular chaperones, often referred to as the "guardians of protein folding," are a diverse class of proteins that specialize in ensuring the correct folding of other proteins. These chaperones act as facilitators, escorting the nascent polypeptide chains and preventing their premature aggregation or misfolding. For membrane proteins, whose hydrophobic segments are particularly prone to misfolding in the watery cytoplasm, chaperones play an indispensable role.

One of the most extensively studied chaperone families is the **heat shock proteins (HSPs)**. These proteins are named for their induction in response to cellular stress, such as elevated temperature, which can cause widespread protein misfolding.

HSPs, including Hsp70 and Hsp90, have been shown to assist in the folding of a wide array of proteins, including membrane proteins.

As an example, consider the folding of a complex integral membrane protein like the **cystic fibrosis transmembrane conductance regulator (CFTR)**, which is crucial for chloride ion transport in epithelial cells. CFTR is a large and intricate protein with multiple membrane-spanning domains. Ensuring its correct folding is a formidable challenge. Hsp70, in association with other co-chaperones, interacts with nascent CFTR chains, preventing their aggregation and guiding them through the folding process. This chaperone-mediated assistance is vital for CFTR's proper function and, by extension, for maintaining normal cellular physiology.

While HSPs play a prominent role in general protein folding, membrane proteins benefit from the assistance of chaperones more specifically tailored to their unique needs. One such group is the **translocon-associated protein (TRAP) complex**, found at the ribosome-translocon interface during co-translational membrane protein insertion. TRAP complexes act as folding assistants by securing and stabilizing the nascent membrane protein chain while ensuring its smooth integration into the lipid bilayer.

Furthermore, the **Sec translocon**, responsible for the co-translational insertion of secretory and membrane proteins into the endoplasmic reticulum, has its own set of chaperone proteins, including **BiP (Binding immunoglobulin Protein)**. BiP ensures the fidelity of the translocation process, actively engaging with hydrophobic segments to prevent aggregation and incorrect folding.

Folding Assistance in Protein Trafficking

For many membrane proteins, the journey doesn't end with proper folding; it extends into the intricate world of protein trafficking. Chaperones continue to play pivotal roles in this phase, ensuring that these proteins reach their intended cellular destinations.

Consider the Golgi apparatus, often called the cellular post office. It's responsible for processing, sorting, and packaging newly synthesized membrane proteins for delivery to their final destinations. In this complex sorting process, chaperones like **COPII (Coat Protein Complex II)** play an indispensable role.

The COPII complex consists of multiple proteins, including Sar1, Sec23, Sec24, Sec13, and Sec31, each with specific functions. The role of COPII is to form vesicles that transport newly synthesized proteins from the endoplasmic reticulum to the Golgi apparatus. Sar1, in particular, plays a critical role in this process by initiating vesicle formation. Additionally, Sec23 and Sec24 are responsible for cargo selection, ensuring that only properly folded and assembled proteins are included in the transport vesicles. This process is vital for preventing the trafficking of misfolded or incomplete membrane proteins to the Golgi apparatus and, ultimately, to the cell surface.

Furthermore, **molecular chaperones like Hsp40 and Hsp70** have been implicated in the quality control of membrane protein trafficking. These chaperones assist in the retrotranslocation and degradation of misfolded or misassembled membrane proteins. They ensure that any proteins that fail to meet the required folding standards are not transported to their intended cellular location.

Chaperones in Disease: Misfolding and Aggregation

The importance of chaperones becomes strikingly apparent when considering the role of protein misfolding in various diseases. Misfolded membrane proteins can lead to a range of pathologies, from neurodegenerative diseases to cystic fibrosis. Chaperones play a significant role in these contexts, offering potential therapeutic avenues.

Alzheimer's disease, for instance, is characterized by the aggregation of misfolded proteins in the brain. Molecular chaperones, such as Hsp70, Hsp90, and small heat shock proteins (sHSPs), have been explored as potential therapeutic targets to prevent or reverse protein aggregation. These chaperones assist in the refolding and clearance of misfolded proteins, offering a glimmer of hope in the fight against neurodegenerative diseases.

Cancer and Protein Folding

The intricate interplay between chaperones and membrane protein folding is also relevant in the context of cancer. Dysregulated protein folding can lead to the production of oncoproteins that drive cancer progression. Molecular chaperones, particularly Hsp90, have attracted attention as potential targets for cancer therapy.

Hsp90 is overexpressed in many cancer types, and it stabilizes various oncoproteins, including kinases and transcription factors, which play critical roles in cancer cell survival and proliferation. Inhibition of Hsp90 has been explored as a strategy to destabilize and degrade these oncoproteins, potentially providing a therapeutic avenue to combat cancer.

Chaperones and folding assistance proteins are the unsung heroes of the intricate world of membrane protein biology. Their indispensable role in guiding nascent polypeptide chains, ensuring their correct folding, and directing them to their intended cellular destinations is a testament to the complexity and precision of cellular processes. Understanding the interplay between chaperones and membrane proteins not only enriches our knowledge of fundamental biology but also holds promise for the development of novel therapeutic strategies in various diseases, from neurodegenerative disorders to cancer. As we delve deeper into the realm of membrane protein research, the significance of these molecular guardians becomes ever more apparent, unravelling new avenues for scientific exploration and medical advancement.

Chapter 12: Membrane Protein Engineering

12.1 Rational Design of Membrane Proteins

In a critical landscape of structural biology, the rational design of membrane proteins stands as a remarkable milestone, offering a precise and strategic approach to engineer these vital cellular components. With the advancement of technology and our deepening understanding of the biological and structural intricacies of membrane proteins, rational design has emerged as a potent tool to modify their properties for specific applications. This section delves into the principles of rational design, exploring the methodologies, applications, and success stories that underscore its profound significance.

Understanding the Rational Design Approach

Rational design, as applied to membrane proteins, entails the targeted modification of protein structures based on a profound understanding of the principles that govern their function. This approach, grounded in the principles of molecular biology and biochemistry, allows researchers to make purposeful alterations to protein sequences, thereby influencing their properties, interactions, and functionalities. The rationale for this approach is rooted in the belief that if we understand a protein's structure and function well enough, we can engineer it to serve specific roles, such as enhanced stability, altered substrate specificity, or modified binding affinities.

Case Study: Rhodopsin Engineering for Optogenetics

One exemplary illustration of rational design in action is the engineering of rhodopsin, a light-sensitive G-protein coupled receptor (GPCR) found in the retina. Rhodopsin is responsible for the initiation of visual processes in response to light. Scientists have harnessed its unique properties to design optogenetic tools, allowing for the control of neuronal activity with light.

In this case, rational design involved the precise alteration of the rhodopsin's amino acid sequence to shift its absorption spectrum towards wavelengths that penetrate tissues more effectively. The result was a new class of rhodopsins known as 'engineered opsins' or 'e-opsins.' These e-opsins have been crucial in the field of optogenetics, enabling non-invasive control of neuronal activity, with profound implications for neuroscience and biomedical research.

Methods in Rational Design

Rational design of membrane proteins requires a comprehensive understanding of the protein's structure and function, often derived from experimental techniques such as X-ray crystallography, NMR spectroscopy, or cryo-electron microscopy. Armed with this structural knowledge, researchers can strategically modify key regions of the protein.

Site-Directed Mutagenesis: This technique involves altering specific amino acids within the protein, typically by replacing them with different amino acids, to investigate their roles in the protein's structure and function. Site-directed mutagenesis is a powerful tool in rational design as it allows for the identification of critical residues and the assessment of their impact on protein properties.

De Novo Design: In some cases, rational design may go beyond simple mutagenesis and extend to the de novo creation of novel proteins with custom functions. Researchers design proteins from scratch, often drawing inspiration from natural protein folds, to achieve specific objectives. This approach is particularly promising for creating membrane proteins with entirely new capabilities.

Computational Methods: Advances in computational techniques have significantly accelerated the rational design process. Molecular modelling and simulation tools enable researchers to predict the impact of sequence modifications on protein structures and functions. Molecular dynamics simulations, in particular, provide insights into how proteins behave in complex lipid environments.

Rosetta Membrane: Rosetta, a widely used software suite, has a membrane-specific module known as Rosetta Membrane, which aids in the design and prediction of membrane protein structures. It offers a valuable platform for structure prediction, protein-protein interaction modelling, and stability assessment, all of which are critical aspects of rational design.

Tailoring Membrane Proteins for Industrial Applications

Rational design of membrane proteins is not limited to the realm of academia and scientific exploration. It has found practical applications across various industries, including pharmaceuticals, biotechnology, and environmental science.

Pharmaceuticals: The pharmaceutical industry has benefited significantly from the rational design of membrane proteins. GPCR engineering, as seen in the development of e-opsins, has paved the way for more targeted drug delivery. By modifying the binding affinities and specificities of GPCRs, researchers can design receptors that respond to new ligands, which has far-reaching implications for drug development.

Biotechnology: In biotechnology, membrane protein engineering has facilitated the creation of tailored receptors for biosensors. These biosensors, designed through rational approaches, can detect specific molecules or ions with high sensitivity and specificity, making them indispensable for applications such as environmental monitoring and medical diagnostics.

Environmental Science: In the field of environmental science, engineered membrane proteins are being used to develop innovative bioremediation techniques. By designing proteins that can selectively bind and remove pollutants from water sources, rational design contributes to cleaner and safer ecosystems.

Challenges and Future Prospects

While rational design has made impressive strides in the engineering of membrane proteins, it is not without its challenges. Predicting the consequences of sequence modifications and their impact on the complex dynamics of membrane proteins remains a formidable task. The structural plasticity of membrane proteins, combined with the intricacies of their lipid environments, can complicate the rational design process.

Future prospects in the field are promising, with ongoing advancements in computational techniques, as well as the growing availability of structural data. These developments will continue to empower researchers to navigate the complexities of membrane protein engineering, opening new doors for applications in medicine, biotechnology, and beyond.

The rational design of membrane proteins is a testament to the remarkable synergy between biology and engineering. This approach, rooted in our ever-expanding knowledge of membrane protein structures and functions, enables us to tailor these proteins to our needs. From designing optogenetic tools for neuroscience to creating biosensors for biotechnology, rational design exemplifies the profound impact of structural biology in addressing real-world challenges. As technology and our understanding of membrane proteins continue to evolve, rational design will undoubtedly remain a central pillar in the quest to harness these essential cellular components for the betterment of society.

12.2 Directed Evolution for Membrane Protein Engineering

Membrane proteins play a pivotal role in various cellular processes, and harnessing their potential for biotechnological and therapeutic applications often requires tailored engineering. Directed evolution, a powerful method, has emerged as a strategic tool for engineering membrane proteins with enhanced functionalities, stability, and altered binding specificities. In this section, we will explore the principles, methods, and notable examples of directed evolution applied to membrane proteins, shedding light on its potential in biotechnological advancements.

Principles of Directed Evolution for Membrane Protein Engineering

Directed evolution is a sophisticated, nature-inspired approach that mimics the process of natural selection in a controlled environment, driving the adaptation of proteins to perform new functions or improve existing ones. The concept underpinning directed evolution involves the generation of genetic diversity through random mutations or recombination, followed by the selection of variants with desired traits. It is a cyclical process, iteratively fine-tuning the properties of proteins. In the context of membrane proteins, directed evolution aims to modify their functions, stability, and interactions with ligands, substrates, or other proteins.

The key steps in directed evolution for membrane protein engineering are as follows:

Mutagenesis: Introducing genetic diversity is a fundamental step in directed evolution. It can be achieved through various methods, such as error-prone PCR, DNA shuffling, or site-directed mutagenesis. These methods induce random mutations in the gene encoding the target membrane protein.

Selection or Screening: After creating a library of mutant proteins, the next step is to identify variants with the desired traits. This can be done through selection or screening. Selection relies on a high-throughput assay to directly isolate mutants showing the desired functionality or property. Screening, on the other hand, involves the testing of individual mutants to identify those with the desired trait.

Amplification and Iteration: Selected or screened mutants are then amplified, and the process is iteratively repeated for several rounds, allowing for the accumulation of beneficial mutations and the continuous improvement of the desired properties.

Characterization: Throughout the process, characterized mutants may undergo detailed biochemical, biophysical, and structural analysis to understand how the directed evolution has affected their properties.

Applications of Directed Evolution in Membrane Protein Engineering

Directed evolution has found a multitude of applications in membrane protein engineering, each aimed at addressing specific challenges or opportunities. Below are some notable examples:

Enhanced Stability: Membrane proteins can be highly sensitive to changes in their environment, which poses a significant challenge for their biotechnological applications. Directed evolution has been used to increase the stability of these proteins under various conditions, including changes in temperature, pH, or ionic strength. For instance, researchers have successfully improved the stability of the G-protein coupled receptor (GPCR), a class of integral membrane proteins, by engineering thermostable variants. These stable variants are valuable in structural studies and drug screening assays.

Altered Specificity: Directed evolution can be employed to modify the binding specificity of membrane proteins. For example, researchers have engineered ligand-gated ion channels to recognize non-natural ligands or to discriminate between closely related compounds. This approach is essential for drug discovery, where selective targeting is crucial.

Functional Enhancement: Some membrane proteins, such as transporters and channels, can be engineered to enhance their transport or conductance capabilities. Directed evolution of membrane transporters, for instance, has led to the development of improved transporters for various substrates, including ions, sugars, and amino acids. These engineered transporters find applications in biotechnology, enabling more efficient nutrient uptake or product secretion in microorganisms used for bioproduction.

Novel Activities: Directed evolution can be used to endow membrane proteins with entirely new functions. In a pioneering example, researchers evolved a GPCR to function as an enzyme, allowing it to activate downstream signalling pathways in response to a non-natural ligand. This concept opens up possibilities for the design of synthetic signalling systems and biotechnological applications.

Methods for Directed Evolution of Membrane Proteins

Directed evolution techniques vary in terms of complexity and specificity. The choice of method often depends on the specific goals of the engineering project. Here are some common methods used for directed evolution of membrane proteins:

Error-Prone PCR: This technique introduces random mutations during the amplification of the gene encoding the membrane protein. The error rate can be controlled by adjusting the conditions, and the resulting library of mutants can be screened or selected for improved properties.

DNA Shuffling: DNA shuffling involves the fragmentation and recombination of related genes from the same or different species. This method is particularly useful when seeking to combine beneficial mutations from different homologous membrane proteins.

Site-Directed Mutagenesis: When the desired mutations are known or well-characterized, site-directed mutagenesis allows for the precise introduction of specific changes in the gene sequence. This method is particularly useful when fine-tuning specific functional properties.

Yeast Display and Phage Display: Display technologies, such as yeast display and phage display, allow for the presentation of membrane proteins on the surface of yeast cells or phage particles, respectively. This enables the selection or screening of membrane protein variants with improved properties.

High-Throughput Screening: Advancements in robotics and automation have made high-throughput screening more accessible. This approach involves rapidly testing a large number of mutants to identify those with the desired traits. It's especially valuable when dealing with extensive mutant libraries.

Notable Examples in Membrane Protein Engineering

Directed evolution has yielded impressive results in the engineering of membrane proteins, with several landmark studies demonstrating its potential. Here are a couple of noteworthy examples:

Engineering of Rhodopsin for Enhanced Stability: Rhodopsin, a GPCR responsible for vision in vertebrates, is a challenging target for structural biology due to its instability. Researchers used directed evolution to improve its stability and solubility. This allowed for the successful crystallization and structural determination of rhodopsin, leading to a deeper understanding of its activation mechanism.

Altering Substrate Specificity of Transporters: Directed evolution has been used to engineer substrate specificity in membrane transporters. For instance, the lactose transporter LacY was evolved to recognize and transport a non-natural substrate, galactoside. This work demonstrated the potential of altering the substrate specificity of membrane transporters for biotechnological applications.

Challenges and Future Directions

While directed evolution has made substantial progress in the field of membrane protein engineering, it is not without challenges. One of the main challenges is the identification of appropriate high-throughput assays for screening or selection. Additionally, the integration of computational methods for predicting the effects of mutations on membrane protein structure and function is an active area of research.

Looking forward, directed evolution, in combination with advances in structural biology techniques, is expected to play a pivotal role in the development of novel therapeutic agents, the design of synthetic biological systems, and the engineering of membrane proteins for industrial applications. With an increasing understanding of membrane protein structure and function, the potential for this powerful approach to transform biotechnology and medicine is vast.

Directed evolution offers a transformative approach to engineer membrane proteins for various applications, from biotechnology to drug discovery. By harnessing the principles of genetic diversity and iterative selection, researchers can modify membrane proteins to enhance their stability, change their specificity, improve their function, or even bestow them with novel activities. Notable examples demonstrate the power of this approach, while ongoing challenges and future directions underscore the continued growth of this exciting field. As our understanding of membrane proteins deepens, directed evolution promises to be a key player in shaping the future of biotechnology and medicine.

12.3 Applications of Engineered Membrane Proteins

The field of membrane protein engineering has emerged as a powerful force in biotechnology and drug discovery. Engineered membrane proteins, often referred to as designer proteins, hold the promise of enhancing the functionality, stability, and specificity of these integral components of cellular membranes. The ability to manipulate membrane proteins at the molecular level has far-reaching implications, with applications spanning from pharmaceuticals to renewable energy production. In this section, we explore some of the notable applications of engineered membrane proteins, showcasing their diverse impact on various industries.

Biopharmaceuticals and Drug Development

One of the most significant areas where engineered membrane proteins have found applications is in the development of biopharmaceuticals and drug discovery. These designer proteins are instrumental in both target identification and drug screening. Let's take a closer look at a few examples:

Protein-based Therapeutics: Engineered membrane proteins can serve as therapeutic agents themselves. For instance, G-protein coupled receptors (GPCRs), a class of membrane proteins, play a crucial role in cell signalling and have numerous pharmaceutical applications. By engineering GPCRs, researchers have developed modified receptors with altered ligand specificities. This can lead to novel drugs that target specific receptor subtypes with reduced side effects. For instance, engineered GPCRs have been developed to target specific opioid receptor subtypes, potentially providing a way to manage pain more effectively with fewer side effects.

High-Throughput Screening: Engineered membrane proteins are pivotal in high-throughput screening assays to identify potential drug candidates. By creating designer versions of membrane proteins, researchers can increase their stability, expression levels, and binding affinity, making them more amenable to screening. The optimization of these proteins can lead to more reliable and efficient drug screening platforms. A case in point is the engineered beta-2 adrenergic receptor, a GPCR used in drug discovery. Modifications to this receptor have improved its stability and binding affinity, making it an ideal target for high-throughput screening assays to discover new asthma and cardiovascular drugs.

Biofuel Production

Membrane proteins, particularly transporters and channels, are integral to the transport of molecules across cellular membranes. In the realm of biofuel production, this property is harnessed to engineer more efficient biofuel-producing microorganisms:

Bacterial Engineering: The transportation of biofuel precursors across the cellular membrane is often the rate-limiting step in biofuel production. Membrane protein engineering can be used to optimize transport systems in bacteria like Escherichia coli, which are commonly used in biofuel production. By engineering transport proteins, such as aquaporins, to enhance the uptake of biofuel precursors, researchers have significantly increased the efficiency of biofuel production processes.

Algal Biofuels: Algae are a promising source of biofuels, and membrane proteins are central to the uptake of carbon dioxide and the transport of photosynthates in algae. By modifying the membrane transport proteins in algae, scientists have enhanced their ability to sequester carbon dioxide and produce biofuels more efficiently. This has the potential to significantly impact the renewable energy sector.

Bioremediation

Engineered membrane proteins also play a critical role in bioremediation, the process of using living organisms to remove or neutralize environmental contaminants. Here, the focus is on membrane transporters that can facilitate the uptake and detoxification of pollutants:

Heavy Metal Removal: Membrane proteins can be engineered to enhance the uptake of heavy metals by microorganisms. For example, the introduction of engineered metal transporters in bacteria can significantly improve their capacity to remove heavy metals from contaminated water sources. This technology holds promise for addressing water pollution issues caused by heavy metal contamination.

Pollutant Degradation: Engineered membrane proteins can also assist in the degradation of environmental pollutants. In some cases, pollutant-degrading enzymes can be engineered into the membrane of microorganisms to enhance their capacity to break down contaminants. For instance, certain strains of bacteria have been engineered to express membrane-bound enzymes that degrade recalcitrant pollutants like polychlorinated biphenyls (PCBs) more efficiently.

Electrochemical Energy Conversion

Membrane proteins are central to the functioning of biological electron transport chains and ion gradients, making them highly relevant in the field of electrochemical energy conversion:

Fuel Cells: Engineered membrane proteins are explored for use in biofuel cells. For example, cytochrome c, a protein involved in electron transport, has been engineered to improve its electron transfer capabilities. Incorporating such designer proteins into biofuel cells can enhance their efficiency and potential for power generation.

Photosynthesis-inspired Energy Conversion: Membrane proteins involved in photosynthesis, such as photosystem I and II, have inspired the development of artificial photosynthetic systems for energy conversion. These systems aim to convert sunlight into electricity or chemical fuels. By engineering these membrane proteins, researchers seek to optimize their performance in synthetic systems.

Neuroscience and Neuroprosthetics

In the realm of neuroscience, engineered membrane proteins have significant implications for understanding neural function and developing neuroprosthetic devices:

Optogenetics: Engineered membrane proteins, particularly light-sensitive ion channels like channelrhodopsins and halorhodopsins, have been used in optogenetics. This technique allows researchers to control the activity of specific neurons with light, providing insights into neural circuits and offering therapeutic potential in neurological disorders.

Prosthetic Devices: Membrane protein engineering is also relevant to the development of neuroprosthetic devices. For instance, the engineering of ion channels in artificial synapses can contribute to the creation of more biocompatible and efficient prosthetic devices for individuals with neurological impairments.

Structural Biology and Drug Target Validation

Finally, engineered membrane proteins have been pivotal in advancing structural biology and validating drug targets:

Structure Determination: Engineered membrane proteins, optimized for stability and crystallizability, have facilitated the determination of high-resolution structures. This has opened new avenues for structure-based drug design. For example, the engineering of beta-adrenergic receptors led to the determination of their crystal structures, providing valuable insights into the development of cardiovascular drugs.

Drug Target Validation: Designer membrane proteins have aided in validating drug targets. By engineering proteins to mimic disease-related mutations, researchers can study their effects and potential drug interactions, advancing our understanding of diseases like cystic fibrosis and certain cancers.

Engineered membrane proteins have transcended the realm of academic research and become vital tools in diverse industries. Their applications in biopharmaceuticals, biofuel production, bioremediation, electrochemical energy conversion, neuroscience, and structural biology are indicative of the significant impact they have on the advancement of science and technology. As we continue to refine our ability to manipulate and optimize these integral components of cellular membranes, the potential for innovation and discovery in these fields remains vast. Membrane protein engineering stands as a testament to the ingenuity of scientists and their capacity to transform fundamental research into practical solutions.

12.4 Future Prospects in Membrane Protein Engineering

As we journey through the ever-expanding landscape of membrane protein research, the prospects for the field of membrane protein engineering appear to be as promising as they are fascinating. The ability to modify, optimize, and even construct membrane proteins with tailored functionalities is at the forefront of scientific innovation, holding significant implications for biotechnology, drug development, and our understanding of biological systems. In this section, we shall explore the future horizons of membrane protein engineering, touching upon exciting developments, technological advances, and their applications in diverse areas.

Designing Membrane Proteins for Therapeutics

One of the most captivating applications of membrane protein engineering is its potential to revolutionize drug discovery and development. Over the past few decades, advances in structural biology have provided us with detailed insights into the architecture of membrane proteins, including G-protein coupled receptors (GPCRs) and ion channels, which are key players in signal transduction and neurotransmission. By manipulating the structural and functional attributes of these proteins, researchers can devise innovative therapeutic strategies.

Take, for example, the GPCRs, a class of membrane proteins that regulate a myriad of physiological processes and are targeted by over 30% of currently prescribed drugs. Recent developments in membrane protein engineering have enabled the creation of designer GPCRs with altered ligand-binding specificities and signalling profiles. This opens up exciting possibilities for tailoring drugs to individual patients, ultimately leading to more precise and effective treatments.

Artificial Photosynthesis and Renewable Energy

Membrane proteins are not confined to the boundaries of cellular biology; they extend their reach into the realm of renewable energy and artificial photosynthesis. In natural photosynthesis, membrane proteins such as photosystem I and II play a vital role in converting solar energy into chemical energy. Researchers are actively engineering these proteins to enhance their efficiency and stability for harnessing solar power.

The concept of biohybrid or synthetic systems, where membrane proteins are integrated into artificial materials, is a particularly promising avenue. By combining natural and synthetic components, scientists aim to develop efficient, sustainable energy sources. These engineered systems have the potential to convert sunlight into chemical fuels or electricity, offering an environmentally friendly alternative to fossil fuels.

Expanding the Toolbox of Molecular Biologists

Membrane protein engineering is not solely focused on applications in biotechnology and energy. It also has a substantial impact on basic research in the life sciences. The ability to engineer membrane proteins has led to the creation of novel tools for molecular biologists. For instance, genetically encoded sensors and optogenetic tools, which allow the manipulation of cellular processes with light, are often built upon engineered membrane proteins.

An illustrative example is the development of channelrhodopsins, light sensitive ion channels derived from microbial rhodopsins. By introducing specific mutations into these proteins, researchers have fine-tuned their properties to precisely control neuronal activity in optogenetics. This has revolutionized the field of neuroscience, enabling the study of neural circuits and the development of potential therapies for neurological disorders.

Understanding Disease Mechanisms

Engineering membrane proteins can offer a deeper understanding of disease mechanisms, particularly in the context of inherited genetic disorders. Cystic fibrosis, for instance, is caused by mutations in the cystic fibrosis transmembrane conductance regulator (CFTR), a chloride ion channel in the cell membrane. By engineering CFTR variants, researchers can simulate disease-related mutations and investigate their impact on channel function.

This approach provides critical insights into the molecular basis of diseases, potentially leading to the development of targeted therapies. Membrane protein engineering not only helps us comprehend the underlying mechanisms but also facilitates drug screening and personalized medicine for individuals affected by such genetic disorders.

Structural Insights and Rational Design

The continued progress in membrane protein engineering is closely intertwined with structural biology. High-resolution structures of engineered membrane proteins offer valuable insights into the impact of mutations and modifications on protein conformation and function. This information, in turn, aids in the rational design of membrane proteins for specific applications.

For instance, the development of ligand-binding pockets in synthetic proteins can be guided by knowledge of natural protein structures. As computational tools become increasingly sophisticated, researchers can simulate the effects of mutations and assess the stability and activity of engineered membrane proteins. This not only accelerates the design process but also reduces the need for extensive trial-and-error experiments.

Challenges on the Horizon

While the prospects of membrane protein engineering are undeniably exciting, several challenges loom on the horizon. One of the primary challenges is the need for more robust and efficient methods for membrane protein expression, purification, and crystallization. These steps are often laborious and time-consuming, limiting the scalability of membrane protein engineering projects.

Another hurdle is the accurate prediction of membrane protein structures and dynamics. Despite significant progress, the field of computational biology still struggles to predict the behaviour of complex membrane proteins with the same precision as soluble proteins. Overcoming these challenges will require interdisciplinary collaboration between structural biologists, computational scientists, and biotechnologists.

Furthermore, ethical considerations regarding the modification and creation of membrane proteins for specific applications must be addressed. As the field advances, a thoughtful and responsible approach to the ethical implications of membrane protein engineering is of utmost importance.

Membrane protein engineering is a field brimming with potential, offering innovative solutions to some of the most pressing challenges in biotechnology, medicine, and renewable energy. The ability to design, modify, and construct membrane proteins with tailored functionalities is poised to shape the future of scientific and industrial attempts.

As we navigate the uncharted territories of membrane protein engineering, researchers must remain vigilant, tackling challenges with determination, collaboration, and ethical responsibility. The future promises a needlepoint of opportunities where the engineered membrane proteins hold the key to unlocking novel therapies, sustainable energy sources, and a deeper understanding of life's intricate mechanisms. Through this journey, science stands on the cusp of a new era where the engineered proteins themselves might one day reveal the secrets of the cosmos.

Chapter 13: Membrane Proteins in Health and Disease

13.1 Role of Membrane Proteins in Cellular Health

Membrane proteins, the molecular sentinels of cells, are crucial players in maintaining cellular health. Their diverse functions in various cellular processes ensure the harmonious functioning of biological systems. This chapter explores the multifaceted role of membrane proteins in cellular health, shedding light on how these proteins impact the fundamental aspects of life. From transporting essential nutrients and ions to transmitting vital signals, membrane proteins are the linchpin of cellular well-being.

The Cellular Membrane: Guardian of Cellular Health

At the heart of every living cell, there lies a delicate balance of processes that govern its existence and, more importantly, its well-being. Maintaining this equilibrium is vital for the normal functioning of a cell, and this is where membrane proteins take the centre stage. The cellular membrane, primarily composed of lipids and embedded membrane proteins, serves as the guardian of cellular health. These proteins play a pivotal role in cellular physiology by mediating various transport processes, cell signalling, and structural support.

Membrane Transport Proteins: The Gatekeepers of Nutrient Uptake

One of the primary responsibilities of membrane proteins in maintaining cellular health is the regulation of nutrient uptake. For a cell to thrive, it must acquire essential nutrients from its environment. Membrane transport proteins, including channels and transporters, facilitate this process. These proteins ensure the passage of crucial molecules such as ions, sugars, amino acids, and vitamins across the hydrophobic lipid bilayer. This transport function is essential for energy production, macromolecular synthesis, and general metabolic activities.

For example, glucose transporters, such as the GLUT family, are integral membrane proteins responsible for glucose uptake in various tissues. The dysfunction of these transporters can lead to metabolic disorders like diabetes, where the cell's ability to take in glucose is impaired. Similarly, ion channels like the sodium-potassium pump (Na+/K+ pump) are vital for maintaining the electrochemical balance across the cell membrane, which is crucial for cell excitability and nerve conduction.

Signal Transduction: Membrane Receptors in Cellular Communication

Cellular health is not just about nutrient uptake but also involves the complex machinery of intercellular communication. Membrane receptors, a subclass of membrane proteins, serve as the liaisons for these important conversations. These receptors act as sensors, relaying extracellular signals to the cell's interior and triggering various responses.

For instance, G-protein coupled receptors (GPCRs) are a diverse family of membrane receptors that transduce signals from a wide array of ligands, including neurotransmitters, hormones, and odorants. Their activation initiates a cascade of intracellular events, leading to cellular responses such as muscle contraction, hormone secretion, and sensory perception. Any malfunction in these receptors can lead to severe health consequences, including cardiovascular diseases, neurodegenerative disorders, and cancer.

Structural Proteins: Maintaining Cellular Architecture

The physical integrity of the cell is another aspect of cellular health that membrane proteins influence. Structural membrane proteins, such as spectrin in red blood cells and integrins in cell adhesion, play a crucial role in maintaining cell shape, stability, and tissue structure. These proteins also contribute to intracellular organization and provide the scaffolding for various cellular functions.

For example, integrins, a group of transmembrane receptors, connect the cell's cytoskeleton to the extracellular matrix, allowing cells to adhere and interact with their surroundings. This adhesion is essential for tissue integrity, wound healing, and immune responses. Disruptions in the function of structural membrane proteins can lead to conditions like muscular dystrophy, where the structural integrity of muscle cells is compromised.

Metabolism and Energy Production

Cellular health is intrinsically linked to the energy supply, which relies on the well-regulated transport of molecules across the cell membrane. Membrane proteins are central to this process.

Mitochondria, the powerhouses of the cell, contain an array of membrane proteins that are integral to the electron transport chain, a critical component of energy production. These proteins facilitate the transfer of electrons, ultimately generating adenosine triphosphate (ATP), the cell's energy currency.

For instance, complex IV of the electron transport chain, also known as cytochrome c oxidase, is a membrane protein that plays a pivotal role in oxygen consumption and ATP production. Dysfunctional membrane proteins in this complex can lead to mitochondrial diseases and energy deficiency disorders, severely impacting cellular health.

Detoxification and Waste Removal

Cells encounter various toxins and waste products that need to be removed promptly to maintain their health. Membrane proteins, particularly those in the liver and kidney cells, play a vital role in this process. Transporters and pumps in these organs are responsible for transporting harmful substances out of the cell and into the bloodstream, where they can be excreted from the body.

For example, P-glycoprotein, a membrane protein encoded by the MDR1 gene, is responsible for pumping out a wide range of foreign compounds, including drugs, from the cells. This protein ensures that these substances do not accumulate within the cell and cause toxicity. Consequently, mutations in the MDR1 gene can result in adverse drug reactions and affect cellular health.

Membrane proteins are the unacknowledged heroes of cellular health, scoring a masterpiece of processes that maintain the equilibrium of life. From nutrient uptake and signal transduction to structural support and waste removal, these proteins are the linchpin of cellular well-being. Understanding their role in health and disease is not only a scientific pursuit but a gateway to potential therapies and interventions for a myriad of human ailments.

13.2 Membrane Proteins in Human Diseases

Membrane proteins play a pivotal role in numerous cellular functions and physiological processes. The intricate network of membrane proteins regulates the transport of ions, the flow of information, and the signalling pathways that govern our biological systems. Any dysfunction in these proteins can lead to severe consequences, including the development of various human diseases. In this section, we will explore the profound impact of membrane proteins in the context of human diseases and how a malfunction in these proteins can trigger pathological conditions.

Understanding the role of membrane proteins in human diseases requires us to fathom the extensive array of disorders associated with their malfunction. These diseases can span from neurological disorders to cancer, and from cardiovascular conditions to rare genetic syndromes. Here, we delve into some illustrative examples, highlighting the significant influence of membrane proteins in the pathogenesis of these maladies.

Cystic Fibrosis: The Defective CFTR Protein

Cystic fibrosis (CF) is a life-altering genetic disorder primarily caused by mutations in the Cystic Fibrosis Transmembrane Conductance Regulator (CFTR) gene. The CFTR protein is an ion channel that spans the cell membrane, with a critical role in maintaining the balance of salt and water across the cell surface in the respiratory and digestive systems. Mutations in the CFTR gene lead to dysfunctional CFTR proteins, causing thick and sticky mucus to accumulate in the airways and pancreas.

This mucosal buildup not only obstructs the air passages, leading to severe respiratory problems, but also impairs the functioning of the pancreas, affecting the digestion and absorption of nutrients. The malfunction of CFTR exemplifies how defects in membrane proteins can result in a complex multi-organ disease like cystic fibrosis.

Alzheimer's Disease: Amyloid Precursor Protein (APP)

Alzheimer's disease is a neurodegenerative disorder that gradually impairs cognitive function and memory. One of the central players in the pathogenesis of Alzheimer's is the Amyloid Precursor Protein (APP), a transmembrane protein. When APP undergoes abnormal cleavage by enzymes known as secretases, it generates amyloid beta (Aβ) peptides, which accumulate in the brain and form plaques, a hallmark of Alzheimer's disease.

The toxic Aβ peptides disrupt neuronal function and ultimately lead to the death of brain cells. While the exact mechanisms behind APP processing and the accumulation of Aβ in Alzheimer's disease remain the subject of ongoing research, the role of membrane proteins in the disorder's development is unquestionable.

Cardiovascular Disease: The Role of Ion Channels

Cardiovascular diseases encompass a wide range of conditions, including hypertension, arrhythmias, and coronary artery disease. Many of these conditions are closely associated with the malfunction of ion channels in cardiac and vascular cells. Ion channels, critical membrane proteins, control the flow of ions such as sodium, potassium, and calcium across cell membranes, regulating the electrical activity of the heart.

For example, mutations in the KCNQ1 gene encoding a potassium ion channel can lead to Long QT syndrome, a disorder characterized by a prolonged QT interval on an electrocardiogram. This condition can result in arrhythmias, fainting, and sudden cardiac death. Such instances underscore the pivotal role of membrane proteins in the functioning of the heart and the development of cardiovascular diseases.

Cancer: The Tale of Receptor Tyrosine Kinases

Cancer is a complex, multifactorial disease, and membrane proteins are significant players in its progression. A prime example is the family of Receptor Tyrosine Kinases (RTKs), including the Epidermal Growth Factor Receptor (EGFR) and the Platelet-Derived Growth Factor Receptor (PDGFR). These transmembrane proteins initiate signalling pathways in response to extracellular signals, promoting cell growth, proliferation, and differentiation.

In cancer, mutations or overexpression of RTKs can lead to uncontrolled cell division and the formation of tumours. For instance, the overexpression of EGFR is associated with various cancers, including lung cancer and glioblastoma. The development of targeted therapies that inhibit RTKs has revolutionized cancer treatment, underscoring the importance of understanding membrane protein function in oncology.

Inherited Genetic Disorders: The Lysosomal Storage Diseases

Lysosomal storage diseases are a group of rare, inherited metabolic disorders, often caused by mutations in genes encoding lysosomal membrane proteins or enzymes responsible for lysosomal function. Lysosomes are membrane-bound organelles involved in the degradation of cellular waste and recycling of biomolecules.

When membrane proteins in lysosomes are defective, cellular waste products accumulate, leading to cellular dysfunction and damage in various tissues. For example, Gaucher's disease results from mutations in the GBA gene, which encodes the lysosomal enzyme glucocerebrosidase. The lack of functional glucocerebrosidase leads to the accumulation of glucocerebroside in macrophages, causing organ damage and a range of symptoms.

These illustrative examples emphasize the diversity of human diseases linked to membrane protein malfunction. The intricate interplay of these proteins in cellular processes makes them key players in the development and progression of various pathologies. Advancements in understanding the structure and function of membrane proteins are critical not only for elucidating disease mechanisms but also for developing targeted therapies to treat or alleviate these conditions. The next section will explore the therapeutic implications of membrane protein research in addressing these diseases.

13.3 *Therapeutic Implications of Membrane Protein Research*

While considering the significance of modern medicine, the pivotal role of membrane proteins is increasingly being unravelled, leading to a profound impact on the development of therapeutic interventions. Membrane proteins, which traverse the lipid bilayer of cellular membranes, are at the heart of various physiological processes, and their malfunction often underlies a multitude of diseases. In this section, we delve into the therapeutic implications of membrane protein research, examining how the insights gained from the structural and functional elucidation of these proteins are catalysing breakthroughs in drug discovery and the treatment of various health conditions.

The Significance of Membrane Proteins in Drug Development

To comprehend the therapeutic relevance of membrane proteins, it is imperative to acknowledge their prevalence as drug targets. A significant fraction of approved pharmaceuticals targets membrane proteins, attesting to their pivotal role in drug development. For example, G-protein coupled receptors (GPCRs), a class of integral membrane proteins, are among the most druggable protein targets, with approximately 30% of drugs on the market acting on them. Notable examples include beta-blockers, which target adrenergic receptors and are widely employed in cardiovascular disease management, and antipsychotics that modulate dopamine receptors.

Another example lies in ion channels, a diverse group of membrane proteins involved in the regulation of various physiological processes, including the nervous and cardiovascular systems. Sodium channel blockers like Lidocaine are used for local anaesthesia, while calcium channel blockers such as Amlodipine are critical in managing hypertension. These cases underscore the substantial impact membrane protein research has had on drug development.

Structure-Guided Drug Design

One of the most remarkable achievements stemming from membrane protein research is the emergence of structure-guided drug design. The elucidation of high-resolution structures of membrane proteins, particularly through X-ray crystallography and cryo-electron microscopy, has provided invaluable insights for rational drug design. The three-dimensional structures of these proteins offer a detailed view of their binding sites, enabling the design of molecules that can interact with the protein in a highly specific manner.

For instance, the crystal structure of the beta-2 adrenergic receptor, a GPCR, provided a structural basis for the development of highly selective drugs for asthma and chronic obstructive pulmonary disease (COPD). By understanding the precise conformation of the receptor, pharmaceutical researchers were able to design drugs that specifically target it, leading to improved patient outcomes and reduced side effects.

Similarly, in the case of the human epidermal growth factor receptor 2 (HER2), which is implicated in various cancers, the availability of its crystal structure has facilitated the design of targeted therapies like Trastuzumab (Herceptin). These therapies have revolutionised the treatment of HER2-positive breast cancer by blocking the receptor's signalling pathway, thereby inhibiting cancer growth.

Overcoming Drug Resistance

Membrane proteins are also central players in the challenge of drug resistance. The adaptability of pathogens, such as bacteria and viruses, as well as cancer cells, often renders conventional drugs ineffective over time. However, understanding the structural basis of membrane protein interactions with drugs can aid in the development of strategies to combat drug resistance.

The Human Immunodeficiency Virus (HIV) is an exemplar in this context. The virus targets the CD4 receptor and chemokine receptors on the surface of T cells to gain entry into host cells. Antiretroviral drugs, like Maraviroc, were developed by targeting the CCR5 chemokine receptor and blocking the virus's entry. However, HIV has exhibited resistance to some of these drugs. Structural insights into the virus-receptor interactions have been pivotal in the development of new drugs that can combat this resistance.

Orphan GPCRs and Drug Discovery

Orphan GPCRs, a subset of GPCRs whose endogenous ligands are unknown, represent an exciting frontier in drug discovery. By elucidating the structures and functions of orphan GPCRs, researchers can potentially identify novel drug targets. For example, the GPR68 receptor, whose natural ligands were unidentified for years, was recently found to be sensitive to extracellular pH changes. This discovery opens doors to the development of drugs that target GPR68 for the treatment of diseases linked to pH dysregulation, including certain cancers.

Furthermore, orphan GPCRs like GPR35, GPR55, and GPR119 have gained attention in drug discovery due to their potential involvement in metabolic disorders and inflammation. The structural and functional characterisation of these receptors offers new avenues for the development of therapeutics targeting conditions like obesity, diabetes, and inflammatory diseases.

Personalised Medicine and Membrane Proteins

The era of personalised medicine is another facet where membrane protein research has made substantial inroads. Personalised medicine aims to tailor medical treatment to individual patients based on their genetic makeup, and the role of membrane proteins is pivotal in achieving this goal.

For instance, the application of pharmacogenomics, which studies the influence of genetic variation on drug response, relies heavily on membrane proteins. Variations in the genes encoding membrane proteins can significantly impact drug metabolism and efficacy. The knowledge of an individual's membrane protein genetics can inform clinicians about the most suitable drugs, dosages, and potential adverse reactions, enhancing treatment outcomes and minimizing side effects.

Challenges and Future Prospects

While membrane protein research has yielded remarkable progress in drug development and personalised medicine, it is not without challenges. Many membrane proteins remain recalcitrant to crystallisation or high-resolution structural determination. Overcoming these challenges necessitates the development of new techniques and technologies, such as nanodiscs and lipidic cubic phase crystallisation, which hold promise for the structural characterisation of currently challenging targets.

The therapeutic implications of membrane protein research are vast and transformative. From the development of drugs targeting GPCRs and ion channels to the design of therapies against cancer and infectious diseases, membrane protein research has revolutionised the pharmaceutical industry. It has enabled the rational design of drugs, offered solutions to combat drug resistance, unveiled the potential of orphan GPCRs, and paved the way for personalised medicine. As our understanding of these crucial proteins continues to expand, we can anticipate even more breakthroughs in healthcare and an increasingly tailored approach to medical treatment, improving the lives of countless individuals.

13.4 Case Studies in Disease-Related Membrane Proteins

In our exploration of the structural biology of membrane proteins, it is imperative to turn our focus towards the profound implications of membrane proteins in human health and disease. Over the years, research in this domain has unveiled an intricate interplay between the structural aspects of these proteins and the development of a myriad of diseases. This section will delve into several case studies to exemplify how the study of disease-related membrane proteins has not only expanded our understanding of pathophysiology but also paved the way for innovative therapeutic strategies.

Cystic Fibrosis Transmembrane Conductance Regulator (CFTR)

Cystic fibrosis (CF), a genetic disorder affecting the respiratory, digestive, and reproductive systems, is primarily attributed to mutations in the CFTR gene. CFTR is an ion channel that regulates the flow of chloride ions across epithelial cell membranes, crucial for maintaining the thin mucus layer in the airways. Mutations in CFTR result in the thickening of mucus, making it difficult for individuals with CF to clear their airways, leading to chronic infections and respiratory complications.

Structural insights into CFTR have been pivotal in understanding the disease mechanism. The first breakthrough came in 2009 when researchers unveiled the 3D structure of a part of the CFTR protein using X-ray crystallography. This structure revealed the pathogenic mutation sites, offering a basis for targeted drug design. Subsequent cryo-EM studies in 2017 provided a high-resolution structure of the entire CFTR protein in its closed and open states, elucidating the conformational changes associated with its function. Such details are essential for the development of modulator drugs, like Ivacaftor, which correct specific mutations and improve CFTR function, significantly enhancing the quality of life for some CF patients.

Beta-Amyloid Precursor Protein (APP) and Alzheimer's Disease

Alzheimer's disease, a neurodegenerative disorder, is characterised by the accumulation of β-amyloid plaques in the brain. These plaques are formed from the cleavage of the amyloid precursor protein (APP) by proteolytic enzymes, and the γ-secretase enzyme, particularly the presenilin subunit, plays a crucial role. Understanding the structure and function of presenilin, a multi-pass membrane protein, is pivotal in Alzheimer's research.

Cryo-electron microscopy studies have unveiled the high-resolution structure of the γ-secretase complex. This knowledge aids in understanding the mechanism of APP cleavage and the rationale for developing γ-secretase inhibitors as potential therapeutics. By targeting the presenilin subunit with inhibitors, researchers aim to reduce the production of β-amyloid, thus potentially slowing the progression of Alzheimer's disease.

G-Protein Coupled Receptors (GPCRs) and Cancer

GPCRs, a class of integral membrane proteins, are involved in a multitude of cellular processes and are implicated in several diseases, including cancer. One striking example is the Epidermal Growth Factor Receptor (EGFR), a GPCR-like protein, which plays a significant role in the regulation of cell growth and proliferation. Aberrant activation of EGFR is a common feature in various cancers.

The crystal structure of EGFR in complex with the cancer drug Erlotinib has provided key insights into how small molecules can bind and inhibit the receptor's tyrosine kinase activity. These structural studies have paved the way for the development of targeted therapies like Gefitinib and Erlotinib, which have shown remarkable success in treating non-small cell lung cancer, particularly in patients with EGFR mutations. Such precision medicine approaches, guided by structural insights, have transformed cancer treatment and improved patient outcomes.

Sickle Cell Anaemia and the Rhodium Proteins

Sickle cell anaemia, a hereditary haemoglobin disorder, results from a single-point mutation in the haemoglobin gene, leading to the production of abnormal haemoglobin, HbS. These abnormal haemoglobin molecules cluster within red blood cells, distorting their shape and impairing their function. However, the underlying membrane protein responsible for this clustering was unknown for decades.

Recent breakthroughs in cryo-EM have allowed the elucidation of the structural basis for HbS clustering. Researchers identified the membrane protein, Band 3, as a key player in this process. Band 3 stabilizes the attachment of abnormal haemoglobin, promoting the formation of clusters. This newfound understanding of the role of Band 3 in sickle cell anaemia may open new avenues for therapeutic interventions aimed at disrupting the interaction between Band 3 and HbS.

Lysosomal Storage Diseases and Lysosomal Membrane Proteins

Lysosomal storage diseases (LSDs) encompass a group of rare genetic disorders in which enzymes responsible for breaking down cellular waste materials are deficient. One striking example is Gaucher's disease, caused by mutations in the GBA gene encoding the lysosomal enzyme glucocerebrosidase. The buildup of glucocerebroside in cells leads to a range of symptoms, including hepatosplenomegaly and neurological issues.

Structural studies of lysosomal membrane proteins, particularly those involved in the transport of substrates, have provided insights into the targeting and trafficking of lysosomal enzymes. This knowledge is crucial for developing enzyme replacement therapies and small molecule chaperones that can enhance the trafficking and stability of mutant enzymes, thereby offering potential treatments for LSDs like Gaucher's disease.

In each of these case studies, the structural elucidation of membrane proteins has been instrumental in uncovering the molecular mechanisms underlying diseases and in the development of targeted therapies. These examples underscore the profound impact of structural biology in advancing our understanding of the pathophysiology of various conditions and in shaping the landscape of modern medicine. They serve as compelling evidence of the powerful synergy between structural biology and translational research, highlighting the potential for innovative therapeutic approaches in the years to come.

Chapter 14: Membrane Protein Bioinformatics

14.1 Sequence Analysis of Membrane Proteins

Sequence analysis of membrane proteins is a fundamental aspect of structural biology, forming the basis for understanding the structural and functional intricacies of these vital cellular components. In this section, we will explore the methods, tools, and significance of sequence analysis in studying membrane proteins, with a focus on practical examples that illustrate its relevance in contemporary research.

The Genomic Era and Sequence Databases

The advent of the genomic era brought about an explosion in the availability of sequence data, enabling researchers to delve into the genetic blueprints of organisms, including humans. This wealth of data has fostered an era of sequence-based analysis, including the comprehensive study of membrane proteins. Central to this pursuit are the vast sequence databases, which house the sequences of various membrane proteins and provide a valuable resource for bioinformatic analysis.

One prominent example is the UniProt database, which offers a comprehensive collection of protein sequences, including detailed annotations and classifications of membrane proteins. UniProt's annotation of membrane proteins identifies them as such and often provides insights into their subcellular localization, topology, and functional domains. This resource is invaluable for researchers seeking to identify and classify membrane proteins from various organisms.

Topology Prediction

Determining the topology of a membrane protein, which refers to the arrangement of transmembrane helices and loops, is a critical step in understanding its structure and function. Bioinformatic tools play a pivotal role in this process, assisting researchers in predicting the transmembrane segments and their orientation within the lipid bilayer.

One commonly used tool for topology prediction is TMHMM (TransMembrane prediction using Hidden Markov Models), which employs statistical models to identify transmembrane regions within a protein sequence. Another widely utilized tool is TOPCONS, which combines multiple prediction methods to offer a consensus prediction, often increasing accuracy.

For example, consider the G-protein coupled receptor (GPCR) family, a class of membrane proteins with diverse functions. Topology prediction tools have been employed to identify the number and location of transmembrane helices in GPCRs, which is crucial for understanding their ligand-binding domains and intracellular signalling pathways.

Multiple Sequence Alignment

Multiple sequence alignment (MSA) is a fundamental technique in sequence analysis that allows researchers to compare and contrast sequences of related membrane proteins. By aligning multiple sequences, commonalities and variations become apparent, shedding light on conserved domains and evolutionary relationships.

One notable example is the comparison of voltage-gated ion channels. These proteins, which are crucial for the electrical excitability of neurons and muscle cells, exhibit sequence conservation in regions responsible for ion selectivity and voltage sensing. Through MSA, researchers can pinpoint these conserved regions, which are essential for their function, and identify sequence variations associated with specific channel types.

Functional Motif Identification

Functional motifs are short, conserved sequences within a protein that often play a critical role in its function. Bioinformatic tools can be employed to identify these motifs, offering insights into the functional significance of specific regions within a membrane protein.

For instance, let's consider the family of ATP-binding cassette (ABC) transporters, which are membrane proteins involved in the active transport of various molecules. Bioinformatic analyses have identified characteristic Walker A and Walker B motifs within the nucleotide-binding domains of ABC transporters. These motifs are essential for ATP binding and hydrolysis, underlining their functional significance.

Phylogenetic Analysis

Phylogenetic analysis provides a means to explore the evolutionary relationships among membrane proteins. By constructing phylogenetic trees based on sequence data, researchers can infer the divergence and relatedness of different protein families.

One illustrative case is the analysis of aquaporins, a family of membrane proteins responsible for water transport. Phylogenetic analysis has revealed the evolutionary history of aquaporins, distinguishing between subfamilies involved in different transport functions. This knowledge aids in understanding the adaptation of these proteins to various physiological roles.

Studying Mutations and Disease-Related Variants

Sequence analysis is also instrumental in understanding the impact of mutations and genetic variants in membrane proteins, especially in the context of human diseases. By comparing the wild-type and variant sequences, researchers can discern changes in amino acids that may lead to altered protein function or disease susceptibility.

For instance, the cystic fibrosis transmembrane conductance regulator (CFTR), a chloride channel in the cell membrane, is associated with various disease-causing mutations. Sequence analysis has enabled the identification of specific mutations in CFTR that lead to cystic fibrosis, a life-threatening genetic disorder. Understanding these mutations is crucial for developing targeted therapies.

Sequence analysis of membrane proteins is a pivotal component of structural biology and bioinformatics. It not only aids in the identification and classification of membrane proteins but also offers insights into their topology, function, and evolutionary history. By studying sequence data, researchers can unravel the genetic basis of membrane protein-related diseases and develop targeted therapies. In the following sections, we will explore other aspects of bioinformatics in membrane protein research, including structural prediction and computational studies of protein dynamics.

14.2 Structural Prediction and Modelling

In pursuit to understand the complex world of membrane proteins, structural prediction and modelling stand as indispensable tools that bridge the gap between sequence data and three-dimensional structural insights. Membrane proteins, with their complex topology, demand innovative computational approaches to decipher their structure, dynamics, and function. This section dives into the fascinating realm of structural prediction and modelling, illuminating the key concepts, methods, and significant contributions in this critical field.

Introduction to Structural Prediction and Modelling

The task of unveiling the three-dimensional architecture of membrane proteins is akin to solving a multifaceted puzzle with numerous missing pieces. Experimental methods like X-ray crystallography and cryo-electron microscopy have undoubtedly yielded remarkable insights, yet they often encounter formidable challenges in the case of membrane proteins. Structural prediction and modelling, however, offer a complementary avenue, promising to unveil the structural needlepoint of these proteins when experimental data is scarce or elusive.

Homology Modelling: Building on Shared Blueprints

One of the most fundamental and widely used methods for structural prediction is homology modelling, also known as comparative modelling. The principle behind this technique is simple: if a protein of known structure shares a substantial sequence similarity with the target membrane protein, it is likely that their structures will exhibit significant similarities as well. The concept here hinges on the age-old adage that form follows function, and evolution preserves structural blueprints in homologous proteins.

Consider, for instance, the G-protein coupled receptor (GPCR) family, a crucial class of membrane proteins involved in signal transduction. With their intricate helical bundle architecture, GPCRs are notoriously challenging to crystallize. Homology modelling has thus been instrumental in elucidating the structures of various GPCRs. For example, the crystal structure of bovine rhodopsin, the first GPCR to be crystallized, paved the way for homology models of numerous other GPCRs, greatly enhancing our understanding of their function.

Template Selection and Sequence Alignment

Homology modelling begins with the selection of an appropriate template protein, which serves as the structural prototype for the target membrane protein. A key criterion for template selection is sequence identity or similarity; the more closely related the sequences, the more reliable the model. However, it's crucial to consider not only the sequence but also the overall structural relevance of the template.

Consider the β2-adrenergic receptor, another GPCR. For its structural elucidation, the template chosen was bovine rhodopsin, a distant relative with approximately 20% sequence identity. While this might seem modest, their shared transmembrane helical bundle architecture justified the choice. Rigorous sequence alignment methods are then applied to align the target and template sequences, identifying equivalent residues and gaps, which will guide subsequent model building.

Model Construction and Refinement

Once the template and target sequences are aligned, model construction begins. Here, the three-dimensional coordinates of the target protein's atoms are calculated based on the corresponding atoms of the template structure. This involves the generation of main chain and side chain coordinates and the adjustment of bond lengths and angles to satisfy geometric constraints. Iterative refinement processes aim to eliminate steric clashes and optimize the model's overall geometry.

To ensure the reliability of the homology model, numerous tools and algorithms are employed. Molecular dynamics simulations, for instance, can refine the model's energetics and address structural flexibility. These simulations apply forces to atoms to mimic their movements over time, providing a more realistic representation of the protein's behaviour in its native environment.

It's worth noting that the accuracy of homology modelling strongly depends on the quality of the sequence alignment and the structural similarity between the target and template. A close structural match, as well as high sequence identity, greatly enhances the reliability of the resultant model. Conversely, when dealing with more distant relatives, modelling accuracy may diminish, necessitating caution and further validation.

Ab Initio Modelling: Unveiling the Unknown

While homology modelling is invaluable when suitable templates exist, the more challenging aspect of membrane protein structural prediction lies in cases where no closely related homologs have been crystallized. In such instances, researchers turn to ab initio modelling, which involves constructing a model from scratch, without the guidance of a template.

Ab initio modelling, often referred to as de novo modelling, is a daunting but rewarding attempt. It requires complex computational algorithms and considerable computational resources to explore the vast conformational space that a protein can assume. The approach begins with the generation of a protein model based on the target protein's primary structure, typically using simplified representations of the protein and the surrounding solvent environment.

Consider, for example, the prediction of the structure of a novel membrane protein implicated in a specific disease. With no closely related templates available, ab initio modelling becomes the go-to strategy. Here, the sequence serves as a starting point, and various physical and chemical constraints, including energy potentials and knowledge of secondary structure propensities, guide the conformational search. Advanced molecular dynamics simulations are employed to sample the vast landscape of possible structures and identify the most energetically favourable ones.

Hybrid Approaches: Combining Homology and Ab Initio Modelling

In many cases, the boundary between homology modelling and ab initio modelling is not clear-cut. Hybrid modelling approaches leverage both techniques to generate more accurate models. This involves incorporating structural information from homologous proteins where available and employing ab initio techniques to predict missing portions of the structure.

Consider a transmembrane helical protein with a well-defined extracellular domain but a less-defined intracellular region. In such cases, the extracellular domain may be modelled based on homologous templates, while the intracellular region, lacking suitable templates, is subjected to ab initio modelling. The resultant hybrid model is a composite that provides valuable structural insights.

Structure Validation and Quality Assessment

Regardless of the modelling approach employed, the validation of the generated models is of paramount importance. Validation involves a series of checks to assess the quality, accuracy, and reliability of the predicted structures. Researchers rely on a battery of metrics and tools to evaluate the models, and these assessments help in determining the robustness of the predictions.

One widely used metric is the Ramachandran plot, which assesses the backbone torsional angles (phi and psi) of amino acids in the protein. Inaccurate torsional angles result in steric clashes and distorted geometry. A good model will exhibit a high percentage of phi-psi angles in the most favoured regions of the plot.

Another essential quality metric is the root mean square deviation (RMSD), which quantifies the structural differences between the model and experimental structures if available. RMSD values below a certain threshold indicate a good fit between the model and the real structure.

Furthermore, validation tools like MolProbity and PROCHECK assess parameters such as bond lengths, bond angles, and atomic clashes. Such assessments collectively contribute to the overall confidence in the model's accuracy.

Challenges and Future Directions

The field of structural prediction and modelling faces persistent challenges, including the accurate prediction of protein-lipid interactions, solvent accessibility, and dynamics within the lipid bilayer. Moreover, improving ab initio modelling for proteins with little or no homologous sequences remains a prominent goal.

Exciting advancements in machine learning and deep learning approaches have begun to play a crucial role in enhancing the accuracy of structural predictions. These techniques leverage vast datasets of experimentally determined protein structures to refine predictions. In particular, deep learning models can learn complex patterns and relationships within protein structures, offering new avenues for improving accuracy.

Another promising avenue is the integration of structural biology techniques, such as NMR and cryo-electron microscopy, with structural prediction and modelling. Integrative approaches that combine data from multiple sources can provide more comprehensive and accurate structural insights.

Thus, structural prediction and modelling represent a cornerstone of membrane protein bioinformatics, enabling researchers to explore the structural landscape of these critical biomolecules. Whether through homology modelling, ab initio techniques, or hybrid approaches, the field continues to evolve, offering fresh insights into the form and function of membrane proteins. As computational methods advance and merge with experimental approaches, our ability to uncover the secrets of membrane proteins becomes increasingly refined and promising.

14.3 Computational Studies of Membrane Protein Dynamics

The structural elucidation of membrane proteins represents a remarkable milestone in our understanding of these biologically critical molecules. However, static structures, as determined by experimental techniques like X-ray crystallography, provide only a snapshot of a protein's conformation. To truly grasp the functional intricacies, one must consider the dynamic nature of membrane proteins. This is where computational studies come into play, offering insights into the motions, fluctuations, and interactions that underpin their functions.

Understanding Dynamics: A Crucial Dimension

The study of membrane protein dynamics explores the motions and changes in conformation that occur on various timescales, ranging from picoseconds to seconds. These dynamics are integral to a protein's function. For example, ion channels open and close to regulate ion flow, transporters undergo conformational changes to shuttle substrates across membranes, and receptors alter their shape to initiate signalling cascades upon ligand binding.

One remarkable aspect of membrane proteins is their adaptability. They are not rigid entities; rather, they are dynamic structures that respond to their environment and cellular cues. Consider G-protein coupled receptors (GPCRs), a family of membrane proteins involved in cell signalling. The activation of a GPCR by a ligand triggers a cascade of conformational changes, leading to downstream signalling events. Understanding these dynamics is vital for drug design, as drugs often target specific receptor conformations associated with particular functions or signalling pathways.

Computational Techniques for Studying Membrane Protein Dynamics

Computational studies of membrane protein dynamics rely on a range of techniques and methods, each tailored to specific timescales and levels of detail. Below, we explore several key approaches:

Molecular Dynamics Simulations (MD): Molecular dynamics simulations are a cornerstone of computational biology. They involve numerically solving Newton's equations of motion for a system of atoms or particles, providing a trajectory of the system's behaviour over time. In the case of membrane proteins, MD simulations can reveal how a protein interacts with its lipid environment, undergoes conformational changes, and experiences thermal fluctuations. These simulations can span from femtoseconds to microseconds, capturing a wide range of dynamic events.

For instance, a study on the nicotinic acetylcholine receptor, a ligand-gated ion channel, employed MD simulations to investigate the opening and closing of its ion channel pore. The simulations revealed that the conformational changes required for ion channel gating are highly dynamic and occur on a microsecond timescale.

Coarse-Grained Simulations: To extend the timescales accessible by MD simulations, coarse-grained models simplify the representation of the system. Here, groups of atoms are treated as a single particle, allowing researchers to simulate processes occurring over milliseconds to seconds. While sacrificing atomic-level detail, these models can provide valuable insights into large-scale conformational changes.

An example involves the study of bacteriorhodopsin, a light-driven proton pump found in archaea. Coarse-grained simulations elucidated the protein's transition from the dark to the light-adapted state, which involves a significant conformational change that pumps protons across the membrane.

Normal Mode Analysis (NMA): NMA is a computational technique that explores a protein's vibrational modes and low-frequency dynamics. It simplifies the analysis by focusing on the harmonic motions of atoms, which can be particularly informative for understanding a protein's global and collective motions.

In the case of the potassium channel, NMA revealed how the channel undergoes subtle vibrations in its closed state, a prelude to the large-scale conformational changes that lead to channel opening upon stimulation.

Enhanced Sampling Methods: Some dynamic events in membrane proteins occur on timescales that are challenging to capture using conventional MD simulations. Enhanced sampling methods, such as umbrella sampling, metadynamics, and replica exchange, provide strategies to accelerate the exploration of specific conformational transitions or rare events.

For example, the activation of the M2 proton channel in the influenza A virus is a rare event. Enhanced sampling methods were crucial in revealing the conformational pathway and the energy landscape of this transition, shedding light on the proton-conducting mechanism.

Network Analysis: Understanding the interactions and communication pathways within a protein is key to uncovering its dynamic behaviour. Network analysis techniques, such as dynamic cross-correlation analysis (DCCA) and community network analysis, help identify key residues and regions involved in signal transmission and allosteric communication.

In a study on the adenosine A2A receptor, a GPCR, network analysis exposed the communication pathways that connect the ligand-binding site to the G-protein binding region, unveiling the allosteric communication essential for receptor activation.

Applications of Computational Studies in Drug Design

One of the most impactful applications of computational studies of membrane protein dynamics is in drug discovery and design. Understanding the dynamic behaviour of membrane proteins can aid in the identification of allosteric sites, the design of ligands that stabilize specific conformations, and the prediction of how drugs may affect a protein's dynamics.

Consider the case of HIV protease, a key drug target for antiretroviral therapy. Computational studies, including MD simulations, helped reveal the dynamic water network within the protease's active site. This knowledge was leveraged to design new inhibitors that exploited the dynamic properties of the protease, resulting in more effective drugs with reduced chances of resistance.

Similarly, for GPCRs, which are a major class of drug targets, the ability to model the receptor's conformational changes upon ligand binding has transformed drug design. It's now possible to predict the binding modes of ligands in different receptor conformations, enabling the development of functionally selective drugs that bias the receptor towards specific signalling pathways.

Challenges and Future Directions

While computational studies of membrane protein dynamics have made significant strides, challenges remain. Accurate force fields, which describe the interactions between atoms in the simulations, are essential for reliable results. Membrane proteins often require specialized force fields due to their interactions with lipid bilayers, which can be challenging to model accurately.

Additionally, the timescales of many biologically relevant events are beyond the reach of current computational methods. Developing even more efficient enhanced sampling techniques and exploring novel approaches will be crucial to address this limitation.

In the future, as computational power continues to advance, we can expect larger and more complex membrane protein systems to be explored. Moreover, machine learning and artificial intelligence will play a growing role in predicting protein dynamics and designing drugs that target specific conformations.

Therefore, computational studies of membrane protein dynamics are integral to our comprehension of these dynamic biomolecules. These studies provide a dynamic dimension to the structural snapshots we gain from experimental methods, offering a holistic view of how these proteins function in their native environment. As our computational tools and techniques advance, so too will our ability to unravel the intricate dance of membrane proteins and harness this knowledge for biomedical applications.

14.4 Databases and Resources for Membrane Protein Research

In the ever-evolving landscape of structural biology, the availability of comprehensive databases and resources is of paramount importance. Membrane protein research, a field replete with intricacies, hinges on such repositories for the dissemination of data, facilitating comparative analyses, and streamlining investigations. In this section, we delve into the realm of databases and resources tailored to the needs of membrane protein researchers, illustrating their significance through the exploration of a few key examples.

The Membrane Protein Data Bank (MPDB)

One of the cornerstone resources in the realm of membrane protein research is the Membrane Protein Data Bank (MPDB). Launched in 2016, the MPDB is a dedicated repository that focuses on the three-dimensional structures of membrane proteins. It offers a user-friendly interface and a comprehensive collection of high-quality structural data.

For instance, the MPDB houses a substantial number of structures elucidated through X-ray crystallography, NMR spectroscopy, and cryo-electron microscopy. Its repository includes various classes of membrane proteins, such as ion channels, transporters, and receptors. This diversity makes it an invaluable resource for researchers interested in investigating the structural intricacies of membrane proteins.

The MPDB not only serves as a repository but also offers a plethora of tools for the structural analysis of membrane proteins. These tools encompass sequence analysis, topology prediction, and structure alignment. Such utilities empower researchers to extract meaningful insights from the stored data, enabling them to make informed decisions in their studies.

PDBTM: A Subset of the Protein Data Bank

Within the broader landscape of protein structural data, the Protein Data Bank (PDB) stands as a paramount resource. However, when it comes to membrane proteins, the PDB Transmembrane Proteins (PDBTM) subset shines as an essential repository. PDBTM, a curated collection within the PDB, is a go-to source for researchers interested in transmembrane proteins.

PDBTM is instrumental in distinguishing between soluble and transmembrane proteins, which can be a challenge given that many structures in the PDB are of soluble proteins. It provides a curated set of transmembrane protein structures, enabling easy access to data that is specifically relevant to the membrane protein research community.

The subset includes structural information on diverse classes of transmembrane proteins, ranging from single-pass receptors to multi-pass transporters. Researchers can readily access structural coordinates, and the associated annotations provide insights into the biological context of these proteins. This facilitates comparative analyses and helps researchers in unravelling the structural underpinnings of membrane protein function.

Orientations of Proteins in Membranes (OPM)

Understanding the orientation of a membrane protein within the lipid bilayer is vital for comprehending its function. The Orientations of Proteins in Membranes (OPM) database plays a pivotal role in this regard. OPM focuses on providing information about the spatial arrangement of membrane proteins within lipid bilayers.

For each membrane protein structure available in OPM, the database furnishes the tilt and rotation angles of the protein with respect to the membrane plane. Moreover, OPM offers the depth of insertion into the lipid bilayer. Such data is essential for investigating the interactions of the protein with lipids, water molecules, and other components of the cellular membrane.

For example, OPM has been instrumental in studying G-protein coupled receptors (GPCRs), a prominent class of membrane proteins. Researchers have used OPM to decipher the orientation of GPCRs in the lipid bilayer, shedding light on their functional mechanisms and ligand binding sites.

The GPCR Database: A Goldmine for GPCR Research

G-protein coupled receptors (GPCRs) are a class of membrane proteins of immense pharmacological importance, as they serve as targets for numerous drugs. The GPCR Database stands as a goldmine for researchers in this field. It provides a comprehensive collection of GPCR structures and related data, making it a central resource for GPCR research.

This database is of particular significance in the drug discovery process. For instance, by accessing the GPCR Database, researchers can study the structures of GPCRs in complex with various ligands. This insight is invaluable for understanding the binding sites and mechanisms of GPCR activation, which, in turn, aids in rational drug design.

The GPCR Database also offers phylogenetic information, allowing researchers to explore the evolutionary relationships among different GPCRs. This is crucial for understanding the functional diversity and commonalities within this protein family.

UniProt: A Comprehensive Protein Knowledgebase

UniProt, a globally recognized resource, is an indispensable asset for membrane protein researchers. This comprehensive protein knowledgebase provides not only sequence data but also a wealth of information about the functions, interactions, and structural features of proteins, including membrane proteins.

For instance, UniProt contains extensive information about membrane protein families and individual members. This includes data on the proteins' subcellular locations, functions, and post-translational modifications. Furthermore, UniProt incorporates cross-references to other relevant databases, enabling researchers to explore a wealth of interconnected data.

UniProt's role extends beyond data retrieval. It offers powerful tools for sequence analysis, allowing researchers to perform sequence alignments and structure predictions. This empowers researchers to explore the relationships between membrane proteins and their soluble counterparts, providing a holistic view of the proteome.

InterPro: Functional Analysis of Membrane Proteins

While structure elucidation is pivotal, understanding the functional aspects of membrane proteins is equally important. InterPro, a database that integrates information from multiple sources, is instrumental in functional analysis.

InterPro employs a suite of predictive models and algorithms to assign functional annotations to protein sequences. For membrane protein researchers, this is particularly useful for predicting domain structures and functional motifs within the protein sequences. By identifying these functional elements, researchers can gain insights into the roles membrane proteins play in cellular processes.

In addition to its analytical tools, InterPro offers valuable information about protein families, domains, and functional sites. This knowledge aids researchers in elucidating the functions of membrane proteins and their evolutionary relationships with other proteins.

Challenges and Prospects

While these databases and resources are invaluable to membrane protein researchers, challenges persist. The field continues to evolve, and keeping databases up-to-date with the latest structures and functional annotations can be a significant challenge. Furthermore, the integration of data from various sources and databases remains a complex task, which requires continuous improvement.

To address these challenges, ongoing efforts are aimed at enhancing data curation, improving data exchange standards, and fostering collaboration among the scientific community. Moreover, advancements in computational methods and artificial intelligence are being harnessed to expedite the analysis of membrane protein data.

Hence, databases and resources are the lifeblood of membrane protein research. They provide a robust foundation for structural and functional investigations, foster collaboration, and enable the research community to harness the wealth of data produced in this dynamic field. As technology and methodology continue to advance, these resources will remain vital for the ever-deepening exploration of membrane proteins and their roles in cellular function and disease. Researchers are encouraged to explore and leverage these valuable assets to advance our understanding of this critical class of proteins.

Chapter 15: Emerging Technologies in Membrane Protein Structural Biology

15.1 Nanodisc Technology

In an ever-evolving scenario of membrane protein structural biology, one technology that has gained significant attention and traction is the use of nanodiscs. These miniature, self-assembling lipid bilayers are revolutionizing the way we study membrane proteins, offering researchers a versatile tool that bridges the gap between the native cellular environment and the in vitro experimental setting. In this section, we will explore the principles, applications, and potential of nanodisc technology, shedding light on its contributions to our understanding of membrane proteins.

Principles of Nanodisc Formation

Nanodiscs are essentially tiny, discoidal lipid bilayers that serve as a scaffold for embedding membrane proteins. The formation of nanodiscs relies on a simple yet ingenious concept: the use of amphipathic membrane scaffold proteins (MSPs). These MSPs, typically derived from apolipoproteins, are water-soluble proteins with hydrophobic and hydrophilic regions. When mixed with lipids and membrane proteins, MSPs self-assemble into small bilayered structures, encapsulating the membrane protein of interest. This self-assembly is driven by the hydrophobic interactions between the MSPs and the lipid acyl chains, effectively "solubilizing" the membrane protein within the nanodisc. The result is a stable, nanoscale disc-shaped lipid bilayer that can be thought of as a surrogate membrane for the embedded protein.

One notable aspect of nanodisc technology is its flexibility. Researchers can tailor the size of nanodiscs by adjusting the molar ratio of MSPs to lipids, making it possible to match the dimensions of the nanodisc to the specific membrane protein being studied. This customization is crucial, as it allows for a more accurate representation of the native environment of the protein in question. Nanodiscs also offer the advantage of incorporating a wide range of lipid compositions, further mimicking the diversity of native membranes and enabling researchers to study proteins in lipid environments that are biologically relevant.

Applications of Nanodisc Technology

Nanodiscs have found applications across a spectrum of membrane protein research, making them a valuable tool in the structural biology arsenal. Here are some key areas in which nanodisc technology has made significant contributions:

Stabilization of Membrane Proteins: One of the most significant challenges in studying membrane proteins is their inherent instability when removed from their native lipid environment. Nanodiscs provide a stabilizing milieu that can prolong the functional lifespan of membrane proteins. This is especially important for structural studies using techniques like X-ray crystallography and cryo-electron microscopy (cryo-EM), which often require purified and well-behaved proteins.

Structural Studies: Nanodiscs serve as an excellent platform for structural determination of membrane proteins. By embedding the protein of interest in a nanodisc, researchers can obtain high-resolution structural information using X-ray crystallography, NMR spectroscopy, or cryo-EM. In recent years, numerous membrane protein structures have been elucidated using nanodiscs, including G-protein-coupled receptors (GPCRs), ion channels, and transporters.

Functional Characterization: Beyond structural studies, nanodiscs facilitate functional assays of membrane proteins. Researchers can investigate the activity and interactions of membrane proteins in a controlled environment, offering insights into their physiological roles and potential drug targets. For instance, electrophysiological measurements of ion channels within nanodiscs have led to a better understanding of their gating mechanisms.

Drug Development: Nanodiscs play a crucial role in drug discovery and development, particularly in the field of membrane protein targets. By incorporating target membrane proteins in nanodiscs, researchers can screen for potential ligands and inhibitors, aiding in the development of new pharmaceuticals. This is particularly relevant for diseases associated with malfunctioning membrane proteins, such as neurodegenerative disorders and certain types of cancer.

Study of Dynamic Processes: The dynamic nature of membrane proteins is central to their function. Nanodiscs allow researchers to investigate these dynamics through techniques like NMR spectroscopy. By capturing the protein in a native-like lipid environment, nanodiscs offer insights into conformational changes, interactions, and kinetics that are challenging to observe using traditional methods.

Recent Breakthroughs and Case Studies

Recent breakthroughs in membrane protein research underscore the impact of nanodisc technology. One remarkable example is the structural elucidation of GPCRs, a class of membrane proteins crucial for signal transduction and targeted by a wide range of pharmaceuticals. GPCRs embedded in nanodiscs have been successfully crystallized, shedding light on their three-dimensional structures and binding sites. This knowledge has far-reaching implications for drug development and our understanding of GPCR signalling pathways.

Another notable case study involves the investigation of membrane transporters. Researchers have employed nanodiscs to study the transport mechanisms of proteins responsible for the movement of ions, metabolites, and drugs across cell membranes. These studies have uncovered crucial details about the conformational changes associated with transport and the binding sites of substrates and inhibitors.

Future Prospects in Nanodisc Technology

Nanodisc technology continues to evolve, with ongoing efforts aimed at enhancing its applicability and versatility. Here are some potential future directions for this technology:

Nanodisc Variants: Researchers are exploring the development of modified nanodiscs that can accommodate larger membrane protein complexes, such as those involved in cellular signalling cascades or multi-protein assemblies. These advances could broaden the scope of proteins that can be studied using nanodisc technology.

Advanced Structural Methods: As cryo-EM and NMR spectroscopy techniques advance, nanodiscs will likely play an increasingly prominent role in these methods. This may lead to higher-resolution structural insights into a broader range of membrane proteins.

Integration with Lipidomics: Understanding the specific lipid composition of nanodiscs and its impact on membrane protein structure and function will be an emerging area of study. This integration with lipidomics will offer a more comprehensive view of the membrane environment.

Industrial Applications: The use of nanodisc technology in industrial settings, particularly in the biopharmaceutical industry, may become more widespread. As the need for structural and functional information on membrane protein drug targets grows, nanodiscs offer a practical solution.

Therefore, nanodisc technology has revolutionized the field of membrane protein structural biology. Its ability to stabilize, mimic native environments, and facilitate a range of experiments has made it an indispensable tool for researchers. The ongoing development of nanodisc variants and its integration with advanced structural methods promise exciting opportunities for the future, offering new insights into the complex world of membrane proteins. This technology continues to be instrumental in advancing our understanding of membrane proteins and their roles in health and disease.

15.2 Lipidic Cubic Phase for Membrane Protein Studies

While considering the significance of structural biology, the mission to untie the anonymities of membrane proteins remains a focal point of research. These proteins, often referred to as the gatekeepers of the cell, are involved in a plethora of biological processes, making them pivotal targets for both fundamental research and drug development. However, studying membrane proteins presents unique challenges, primarily due to their hydrophobic nature and the necessity to maintain their structural integrity in a physiologically relevant environment. Here, we delve into a groundbreaking technology - the lipidic cubic phase - which has emerged as a game-changer in the study of membrane proteins, offering an exciting avenue for scientists to navigate this challenging terrain.

Introduction to the Lipidic Cubic Phase

The lipidic cubic phase (LCP), sometimes referred to as the bicontinuous cubic phase, is a novel technique that has gained substantial traction in the structural biology community for its remarkable ability to crystallize and study membrane proteins. Membrane proteins are known to reside in the lipid bilayer of cell membranes, surrounded by a hydrophobic environment, which poses a significant challenge when attempting to extract, purify, and crystallize these proteins in a manner that preserves their native structure and function. Traditional methods like detergent solubilization often disrupt the native lipid environment, potentially altering the protein's conformation.

The LCP technique offers a solution by providing a biomimetic environment that mimics the lipid bilayer in which membrane proteins naturally reside. This phase consists of an interconnected network of water channels within a lipid matrix, creating a 3D lattice structure. In this crystalline lipidic matrix, membrane proteins can be embedded, offering an environment that closely resembles their native surroundings. This technique effectively mitigates the disruption of the lipid-protein interactions, thereby preserving the protein's native conformation and functionality.

LCP in Membrane Protein Crystallization

One of the primary applications of the LCP technique is in membrane protein crystallization. Crystallization is a pivotal step in structural biology, as it allows for the determination of the protein's three-dimensional structure using techniques like X-ray crystallography. Membrane proteins, being inherently challenging to crystallize due to their hydrophobic nature and the necessity of maintaining their native lipid environment, have historically been elusive subjects for structural analysis.

LCP's ability to create a lipidic environment in which membrane proteins can naturally exist has revolutionized the field. In LCP, the crystallization process typically begins by mixing a solution of lipid molecules with the purified membrane protein of interest. This mixture forms a cubic phase, entrapping the protein within the lipid lattice. As a result, the protein is preserved in a native-like state, and over time, this phase can transform into crystals suitable for X-ray diffraction studies.

Advantages of LCP in Membrane Protein Studies

The adoption of the LCP technique has introduced several notable advantages for membrane protein studies:

Native-Like Environment: LCP provides an environment that closely mimics the lipid bilayer, allowing membrane proteins to maintain their native conformation, function, and interactions with surrounding lipids. This is particularly crucial when studying the structural details of these proteins.

Enhanced Stability: Membrane proteins are notoriously delicate, and changes in their environment can lead to denaturation and loss of function. LCP helps protect the structural stability of these proteins during the crystallization process.

Crystallization Success: Many membrane proteins are resistant to crystallization using traditional methods. LCP has proven highly successful in crystallizing a wide range of membrane proteins, expanding the scope of what can be structurally characterized.

Detergent-Free: Unlike traditional detergent-based methods, LCP allows researchers to study membrane proteins without the interference of detergents, which can disrupt the protein's natural interactions with lipids.

Improved X-ray Diffraction: LCP-grown crystals often yield high-resolution X-ray diffraction data, which is essential for obtaining detailed structural information.

Applications in Drug Discovery

The structural information obtained from membrane proteins using LCP has profound implications for drug discovery. Many therapeutic targets are membrane proteins, including G-protein-coupled receptors (GPCRs), ion channels, and transporters. Understanding the structure of these proteins in their native environment can aid in the rational design of drugs that target them.

For example, GPCRs, a class of membrane proteins, are involved in a wide range of physiological processes and are one of the most common drug targets. The LCP technique has enabled the crystallization of several GPCRs, providing insights into their activation mechanisms and ligand binding sites. This information has opened doors to the development of more effective and specific drugs targeting these receptors.

Case Studies in LCP Applications

Several notable case studies highlight the impact of the LCP technique in membrane protein research:

β2 Adrenergic Receptor: In 2007, the structure of the β2 adrenergic receptor, a GPCR, was determined using LCP. This breakthrough provided critical insights into the activation of GPCRs, paving the way for the design of drugs targeting these receptors.

Rhodopsin: The structure of rhodopsin, a membrane protein involved in vision, was determined using LCP. This work shed light on the mechanism of phototransduction and the conformational changes that occur when rhodopsin is activated by light.

Bacterial Rhomboid Protease: LCP was instrumental in elucidating the structure of a bacterial rhomboid protease. This protease plays a role in various cellular processes and is a potential target for antimicrobial drug development.

GABA Transporter: The GABA transporter, involved in neurotransmission, was crystallized in an LCP matrix, offering insights into its transport mechanism and potential implications for neurological drug development.

The lipidic cubic phase has emerged as a transformative technology in the field of structural biology, particularly in the study of membrane proteins. By providing a native-like environment for these proteins during crystallization, LCP has overcome many of the challenges associated with traditional methods. This innovative technique has expanded the scope of what we can learn about membrane proteins, offering new opportunities for drug discovery and a deeper understanding of the fundamental biology of these crucial cellular components. As technology continues to advance, the lipidic cubic phase holds promise for even greater insights into the intricate world of membrane proteins and their roles in health and disease.

15.3 Advances in Labelling Techniques

In the hunt to decode the structural details of membrane proteins, labelling techniques have played a pivotal role. These techniques are the brushstrokes in the canvas of structural biology, allowing researchers to add the essential details to the portraits of these enigmatic biomolecules. This section explores the recent advances in labelling techniques, shedding light on how researchers are bringing these proteins into sharper focus.

The journey towards understanding membrane proteins at the molecular level hinges on the ability to label specific sites or segments within these proteins. Labelling, in this context, is akin to adding colour to a grayscale photograph, accentuating key features and allowing for a more detailed examination. The most recent advances in labelling techniques have ushered in a new era of precision and accuracy, enabling researchers to explore membrane proteins with unprecedented clarity.

Site-Specific Labelling

One of the most significant advancements in membrane protein labelling techniques is the move towards site-specific labelling. Traditional labelling methods often resulted in a haphazard distribution of labels, making it challenging to draw meaningful conclusions about the protein's structure and function. Site-specific labelling, however, targets specific amino acid residues within the protein, providing a more refined and accurate view.

For example, the use of genetically encoded unnatural amino acids (UAAs) has revolutionized site-specific labelling. UAAs are incorporated into the protein sequence during expression, introducing unique chemical handles for labelling. This technique allows researchers to attach a wide range of labels, from fluorescent dyes to paramagnetic probes, precisely where they want within the protein. It not only simplifies labelling but also reduces the potential for perturbing the protein's native structure or function.

Click Chemistry

Click chemistry, a term coined by K. B. Sharpless and his colleagues, refers to a set of reactions that are highly selective, generate minimal byproducts, and proceed under mild conditions. This approach has found extensive use in bioconjugation, a key aspect of labelling techniques. By employing click chemistry, researchers can attach labels to specific functional groups, enhancing the precision of labelling.

For instance, the copper-catalysed azide-alkyne cycloaddition (CuAAC) reaction is one of the most widely used click chemistry methods in labelling membrane proteins. This reaction involves the coupling of an azide-tagged molecule with an alkyne-tagged molecule in the presence of a copper catalyst. It is highly selective and biocompatible, making it suitable for labelling studies on live cells. This technique has been instrumental in tracking the dynamic behaviour of membrane proteins within their native environment.

Bioorthogonal Labelling

Bioorthogonal labelling refers to the introduction of non-natural chemical functionalities into biomolecules, allowing for the specific labelling of these molecules without interfering with other cellular processes. The advantage of bioorthogonal labelling is that it can be tailored to precisely target membrane proteins, minimizing off-target effects.

A notable example of bioorthogonal labelling is the use of azide-alkyne cycloaddition reactions, which are copper-free and highly selective. These reactions can be employed in a variety of labelling strategies, including fluorescent labelling, biotinylation, and even the attachment of nanoparticles, enabling the visualization and tracking of membrane proteins in diverse experimental contexts.

Proximity-Dependent Labelling

Proximity-dependent labelling techniques have gained popularity for their ability to unveil protein-protein interactions and the spatial organization of membrane proteins within cellular membranes. These techniques rely on the fusion of the protein of interest with a proximity label, which can be activated under specific conditions.

As an illustration, the biotin ligase-based proximity-dependent labelling system, known as BioID, has been instrumental in mapping protein-protein interactions of membrane proteins. In BioID, the protein of interest is fused to a promiscuous biotin ligase, BirA*, which biotinylates nearby proteins. Biotinylated proteins can then be isolated and identified, offering valuable insights into the membrane protein's interactome.

Fluorescence Resonance Energy Transfer (FRET)

Fluorescence Resonance Energy Transfer, or FRET, has long been a stalwart technique in studying protein-protein interactions and conformational changes. FRET relies on the transfer of energy between a donor and an acceptor fluorophore when they are in close proximity, making it a powerful tool for exploring the dynamic behaviour of membrane proteins.

For instance, FRET has been employed to study the dimerization of G-protein coupled receptors (GPCRs), a class of membrane proteins with critical roles in cellular signalling. By labelling specific sites on GPCRs with donor and acceptor fluorophores, researchers can monitor changes in receptor conformation and interactions with downstream signalling partners, shedding light on the intricate workings of these proteins.

Solid-State NMR and Dynamic Nuclear Polarization (DNP)

Solid-state NMR has emerged as a powerful technique for studying the structure and dynamics of membrane proteins embedded in lipid bilayers. Recent advancements in dynamic nuclear polarization (DNP) have significantly enhanced the sensitivity of solid-state NMR, making it a valuable tool for investigating membrane protein structure and function.

Notably, DNP-enhanced solid-state NMR has been used to probe the structure of amyloid fibrils, which are associated with neurodegenerative diseases. This technique has allowed researchers to gain atomic-level insights into the structure of membrane proteins implicated in these diseases, opening new avenues for drug development.

Nanobodies and Nanoparticles

Nanobodies, single-domain antibodies derived from camelids, have gained recognition for their utility in membrane protein research. These small, stable proteins can be engineered to bind specifically to membrane proteins, facilitating their purification and labelling.

A compelling example is the use of nanobodies in the structural analysis of GPCRs. These small, robust binders can be labelled with fluorescent dyes, making them ideal tools for tracking the conformational changes of GPCRs in response to ligand binding.

Nanoparticles have also made a mark in the field of membrane protein labelling. These tiny structures can carry a variety of labels and can be specifically engineered to interact with membrane proteins, making them valuable tools for both labelling and structural analysis.

Super-Resolution Microscopy

Super-resolution microscopy techniques have transformed our ability to visualize cellular structures, including membrane proteins, at a resolution beyond the diffraction limit of conventional light microscopy. These techniques provide a new level of detail in the study of membrane protein organization and distribution.

For example, stochastic optical reconstruction microscopy (STORM) and photoactivated localization microscopy (PALM) enable the precise localization of individual molecules. Researchers have employed these techniques to elucidate the nanoscale distribution of membrane proteins, uncovering intricate patterns of organization within cellular membranes.

Small-Angle X-ray and Neutron Scattering (SAXS and SANS)

Small-angle X-ray scattering (SAXS) and small-angle neutron scattering (SANS) have become powerful tools for studying the shape and organization of membrane proteins in solution. These techniques provide valuable information about the overall structure and conformation of proteins without the need for crystallization.

In practice, SAXS and SANS have been applied to elucidate the structure of integral membrane proteins, such as ion channels and transporters. By collecting scattering data from these proteins in different solution conditions, researchers can gain insights into their conformational changes and oligomeric states, providing crucial information for drug development and structural biology.

Advances in High-Throughput Labelling

The demands of structural biology have pushed researchers to develop high-throughput labelling techniques, which allow for the rapid and systematic labelling of multiple membrane proteins simultaneously. These advancements have accelerated the pace of structural studies on these proteins.

As an example, the development of automated liquid-handling systems and microfluidic devices has streamlined the labelling process, enabling researchers to investigate large numbers of membrane proteins in a high-throughput manner. This is particularly valuable in drug discovery efforts, where the evaluation of multiple protein targets is essential.

In Vivo Labelling Techniques

The ability to label membrane proteins within living cells is of paramount importance for understanding their dynamic behaviour in their native environment. Recent developments in in vivo labelling techniques have allowed for real-time tracking of these proteins within cells.

For instance, the advent of the CRISPR/Cas9 system has enabled the precise genome editing of cells to express tagged membrane proteins. By fusing these proteins to fluorescent tags or other labels, researchers can follow their localization and interactions in real-time, providing invaluable insights into their function within the cellular context.

Challenges and Future Directions

While these advances in labelling techniques have brought membrane protein research to new heights, challenges remain. Achieving complete site-specific labelling without perturbing protein function, enhancing labelling efficiency, and reducing background noise are areas where continued improvements are needed.

Future developments in labelling techniques may focus on integrating multiple approaches to provide a more comprehensive understanding of membrane proteins. Combining structural techniques, such as cryo-electron microscopy (cryo-EM), with labelling methods could offer a multi-faceted view of these proteins, elucidating their dynamic behaviour, interactions, and structural details in a single study.

The recent advances in labelling techniques have significantly enriched our understanding of membrane proteins. From site-specific labelling to super-resolution microscopy, these techniques are the key to unlocking the secrets of these essential biomolecules. As technology continues to evolve, so too will our ability to paint a clearer picture of membrane protein structure and function, ultimately advancing our knowledge of cellular processes and opening new avenues for drug discovery and therapeutic interventions.

15.4 Future Frontiers in the Field

As we journey through the vast landscape of membrane protein structural biology, it becomes apparent that this scientific terrain is dynamic and ever-changing. This chapter sets its gaze on the horizon, peering into the uncharted waters of future frontiers in the field. In this section, we explore some of the emerging avenues and evolving trends that promise to shape the course of membrane protein research in the years to come.

The field of membrane protein structural biology, like a scientific adventure, never ceases to evolve. It continually ushers in innovative technologies and approaches that push the boundaries of our understanding. As we stand at the threshold of the future, several exciting frontiers beckon us, promising to unveil the mysteries hidden within the membranes of life. Here, we delve into the currents of these future frontiers, each with the potential to redefine our approach and impact.

Membrane Protein Dynamics in Native Environments

Our understanding of membrane protein dynamics has advanced significantly over the years, primarily driven by developments in nuclear magnetic resonance (NMR) and computational techniques. However, the majority of structural studies still rely on snapshots in artificial environments, often devoid of lipids and other key components of the native cellular milieu. The next frontier is to probe the dynamics of membrane proteins in their natural settings.

Recent strides in cryo-electron tomography (cryo-ET) offer a promising avenue in this regard. Cryo-ET enables the three-dimensional visualization of macromolecules within intact cellular membranes. It allows us to witness proteins in action, be it in the lipid bilayer or at membrane interfaces, shedding light on conformational changes, protein-lipid interactions, and functional dynamics in their native context.

To exemplify this frontier, consider recent studies on bacterial chemoreceptors. Cryo-ET has revealed that these receptor clusters are not static but highly dynamic, with receptors forming large, heterogeneous arrays. Understanding such dynamic organization is crucial for deciphering cellular signalling mechanisms and developing targeted therapies.

Integrating Multi-Omics Data for Systems-Level Insights

The advancement of 'omics' technologies has opened new vistas in biology, including the proteome and lipidome. A pressing frontier in membrane protein research is to integrate data from multiple 'omics' domains to acquire a holistic view of the cell's molecular machinery.

For instance, combining lipidomics data with structural information can offer insights into the role of lipids in modulating membrane protein function and localization. This multidisciplinary approach is imperative for unravelling the complexities of lipid-protein interactions, leading to a deeper understanding of cellular processes.

Consider the study of rhodopsin, a prototypical G-protein coupled receptor (GPCR). By merging lipidomics data with cryo-EM structures, researchers have revealed the importance of specific lipid molecules in stabilizing the receptor's active state. These insights can have profound implications in drug design for conditions such as retinal degeneration.

Conformational Landscapes and Allosteric Regulation

Membrane proteins often exhibit intricate conformational landscapes, crucial for their multifaceted functions. The concept of allostery, where a change in one part of the protein induces a change in another, is a central theme in this frontier.

Advanced computational approaches, coupled with high-resolution structural techniques, are beginning to provide a more comprehensive understanding of allosteric sites in membrane proteins. These sites can be targeted for novel drug design strategies.

For example, the bacterial enzyme AcrB, a component of the efflux pump, is a prime candidate for allosteric modulation. By mapping its conformational landscape through molecular dynamics simulations and combining this with structural data, researchers have identified potential allosteric pockets that could be targeted to inhibit antibiotic resistance.

Nanotechnology and Beyond: The Quest for Precision Medicine

The dawn of precision medicine promises a revolution in the treatment of diseases. A key aspect of this revolution is the development of personalized therapeutics targeting membrane proteins. The future frontier in this realm is the amalgamation of membrane protein research with nanotechnology.

Nanotechnology allows for the precise manipulation of materials at the nanoscale, offering innovative solutions for drug delivery and targeted therapy. Gold nanoparticles, for instance, can be functionalized with membrane protein ligands, resulting in highly specific drug carriers that can traverse the blood-brain barrier for the treatment of neurological disorders.

Additionally, the concept of synthetic biology intersects with nanotechnology, giving rise to the prospect of engineering designer membrane proteins tailored for therapeutic purposes. By designing proteins with specific properties and functions, we can open the door to a new era of precision medicine.

The Rise of In Silico Drug Design

In silico drug design, a computational approach to drug discovery, is set to play an increasingly pivotal role in membrane protein research. This frontier leverages computational power to simulate the interactions between drugs and their target membrane proteins, accelerating the drug development process.

An illuminating example is the research on voltage-gated ion channels, integral membrane proteins implicated in numerous diseases. Through computational simulations, researchers can now rapidly screen a vast library of compounds to identify potential drug candidates, reducing the time and cost associated with experimental drug discovery.

The future holds the promise of developing drugs with higher specificity and efficacy, while minimizing the risk of adverse effects, by harnessing the power of in silico drug design. This convergence of computational and experimental approaches is a frontier that is reshaping the landscape of drug discovery.

The future frontiers of membrane protein structural biology are a testament to the field's vibrant and ever-evolving nature. From probing dynamics in native environments to integrating multi-omics data, unravelling allosteric regulation, embracing nanotechnology for precision medicine, and harnessing the potential of in silico drug design, these frontiers hold the keys to unlocking new dimensions in our understanding of membrane proteins. The journey ahead is one of excitement, challenge, and unending discovery, inviting the scientific community to embark on this remarkable voyage into the unknown.

Chapter 16: Ethical Considerations in Membrane Protein Research

16.1 Ethics in Scientific Research

Scientific research stands as a beacon of human curiosity, illuminating the path to discovery, innovation, and progress. It is an attempt characterised by rigorous exploration, driven by the pursuit of knowledge, and underpinned by an unwavering commitment to ethical principles. In this section, we will navigate the critical domain of ethics in scientific research, exploring the foundations, challenges, and evolving paradigms that shape the moral compass of researchers.

The Moral Compass of Scientific Research

Ethics is the guiding star of scientific inquiry, ensuring that the pursuit of knowledge remains resolutely anchored to principles of integrity and responsibility. At its core, ethics in scientific research encompasses a broad spectrum of values, standards, and practices that uphold the credibility, transparency, and trustworthiness of the scientific enterprise.

A fundamental principle that guides ethical research is integrity, which serves as the bedrock of scientific truth. Integrity dictates that researchers must be honest and truthful in all their actions, from data collection and analysis to the dissemination of findings. Maintaining the sanctity of data and results, irrespective of their alignment with initial hypotheses, is a hallmark of scientific integrity. The infamous cases of scientific misconduct, such as the fabricated data in the case of Korean stem cell researcher Hwang Woo-suk in 2005, serve as stark reminders of the perils of compromising scientific integrity.

Moreover, ethics in scientific research extends to the treatment of research subjects and animals. Researchers are bound by an ethical duty to ensure that their research is conducted with the utmost respect for the rights, dignity, and well-being of human subjects. Informed consent, confidentiality, and protection of privacy are non-negotiable principles in any study involving human participants. Similarly, research involving animals demands adherence to the '3Rs' principles: Replacement, Reduction, and Refinement. This framework underscores the ethical imperative to replace animals in experiments where possible, reduce the number of animals used, and refine experimental procedures to minimize suffering.

The Challenge of Balancing Exploration and Responsibility

The pursuit of scientific knowledge is a relentless quest, driven by curiosity, ambition, and a deep-seated desire to unravel the mysteries of the universe. Yet, this unquenchable thirst for discovery is not without its ethical conundrums.

One of the most complex ethical dilemmas in scientific research is the dual role of the researcher as both an explorer and guardian. Researchers are charged with the task of pushing the boundaries of knowledge while simultaneously safeguarding the rights and welfare of those affected by their work. Balancing these often conflicting roles requires a profound commitment to ethical conduct.

For instance, consider the field of genetic research. The mapping of the human genome, a monumental scientific achievement, has opened the doors to a realm of possibilities, from understanding genetic disorders to potential gene therapies. However, this research also raises ethical concerns about privacy, discrimination, and informed consent. Researchers must navigate these treacherous waters with caution, ensuring that while they explore the genetic needlepoint of humanity, they do not violate the ethical principles that protect individuals' rights.

Furthermore, advancements in emerging technologies, such as artificial intelligence and gene editing, present novel ethical challenges. The potential for these technologies to revolutionise fields like medicine and agriculture is undeniable, but the ethical implications of manipulating genetic code or ceding control to autonomous machines must not be ignored.

Emerging Paradigms in Research Ethics

As scientific research advances, so too does the terrain of research ethics. New paradigms are continually evolving, influenced by societal, technological, and cultural shifts. One such paradigm that has gained prominence in recent years is Open Science.

Open Science is a movement that advocates for transparency, collaboration, and the unrestricted sharing of research data and findings. It aims to democratise science by making research more accessible to the broader scientific community and the public. By adhering to Open Science principles, researchers can enhance the rigour and reproducibility of their work, contributing to the overall integrity of the scientific enterprise.

In a parallel shift, citizen science has emerged as an ethical frontier in scientific research. Citizen science projects involve the active participation of non-professional volunteers in scientific research. These projects have led to a proliferation of data and perspectives, enriching scientific inquiry. However, ethical considerations arise in ensuring that participants' contributions are respected and that they receive appropriate acknowledgment and benefits.

Moreover, the ethics of global collaboration are becoming increasingly prominent as scientific research transcends geographical boundaries. Collaborative international research efforts are essential in addressing global challenges such as climate change, pandemics, and sustainable development. Ethical considerations in global collaboration encompass issues of equity, access, and the fair distribution of benefits, all of which necessitate careful negotiation.

Ethical Challenges in Emerging Fields

In the ever-expanding landscape of scientific research, emerging fields often present unique ethical challenges. Let us consider the field of neuroethics as an example. Neuroethics grapples with the ethical implications of advances in neuroscience and neurotechnology. These advances have led to the development of brain-computer interfaces and the potential for cognitive enhancement. Ethical concerns in neuroethics span issues of autonomy, consent, privacy, and the potential for unequal access to cognitive enhancements.

Similarly, the intersection of biotechnology and environmental conservation gives rise to novel ethical dilemmas. Gene editing technologies, such as CRISPR, offer the possibility of mitigating biodiversity loss and addressing ecological challenges. However, ethical concerns encompass the preservation of natural ecosystems, the potential consequences of unintended genetic alterations, and the ethical balance between human intervention and the preservation of the natural world.

In addressing these ethical challenges, researchers and institutions must be proactive in engaging with stakeholders, including ethicists, policymakers, and the public. Transparent dialogue and ethical considerations should be integral to the research process, helping to anticipate, mitigate, and navigate complex ethical dilemmas.

Ethics in scientific research is not a static entity but a dynamic and evolving facet of the scientific enterprise. It is a testament to the commitment of the scientific community to uphold the principles of integrity, responsibility, and transparency while navigating the uncharted waters of discovery. As research continues to shape the future, the ethical compass remains unwavering, guiding researchers towards the higher ideals of knowledge and societal progress.

In the chapters that follow, we will delve deeper into various aspects of structural biology, continually mindful of the ethical considerations that underpin our exploration of the membrane proteins that define life's fundamental processes.

16.2 Animal Research and Alternatives

The study of membrane proteins is an area of research vital for understanding various cellular processes and developing therapeutics. Yet, it poses ethical challenges, particularly in the context of animal research. This section delves into the ethical dilemmas surrounding animal research in membrane protein studies and explores emerging alternatives that have the potential to revolutionize the way we investigate these proteins.

Ethical Dilemmas in Animal Research

Historically, the use of animals in scientific research, including membrane protein studies, has played a significant role in advancing our understanding of biology and medicine. However, this practice is not without its ethical concerns.

Animal Welfare

One of the primary ethical dilemmas revolves around animal welfare. The utilization of animals, particularly mammals like mice, rats, and primates, raises questions about their well-being, treatment, and the moral obligation to minimize any potential harm. Membrane protein research, often involving invasive procedures, poses risks to the animals involved, both in terms of pain and distress. Scientists and ethicists have grappled with finding a balance between scientific progress and animal welfare.

Moral Status of Animals

The moral status of animals is a contentious issue in the ethical discourse surrounding animal research. Some argue that animals have intrinsic value, deserving respect and protection, irrespective of their utility in research. Others contend that the potential benefits to human health justify their use, albeit with appropriate safeguards. This debate reflects a fundamental tension between the duty to alleviate human suffering and the duty to prevent animal suffering.

Species-Specific Considerations

Considering the diversity of animals used in research, it is essential to recognize that not all species are equal in their cognitive capacities, sensitivity to pain, or capacity to experience suffering. For instance, the ethical considerations of using primates in research differ significantly from those involving simpler organisms like zebrafish. Researchers must account for these differences when designing experiments and ethical guidelines.

Alternatives to Animal Research

In response to the ethical dilemmas surrounding animal research in membrane protein studies, scientists have been actively exploring alternative methods to advance their work while reducing reliance on animals.

In Vitro Methods: In vitro (literally "in glass") methods are experiments conducted outside of a living organism. These techniques are essential in membrane protein research, offering various advantages over in vivo animal experiments. For instance, cell cultures enable researchers to study the behaviour of membrane proteins in a controlled environment. They are particularly useful for understanding the biochemical and biophysical properties of these proteins.

Example: Membrane protein expression in human cell lines, such as HEK293, allows for the investigation of protein function and interactions without the need for animal subjects. This approach is especially valuable in G-protein coupled receptor (GPCR) research, as it provides insights into the receptor's signalling pathways and pharmacological responses.

Organoids and Tissue Chips: Advancements in tissue engineering have given rise to organoids and microphysiological systems, often referred to as "tissue chips." These technologies enable the creation of three-dimensional, miniature organs or tissues from human cells. Organoids and tissue chips serve as valuable models for studying the effects of drugs or diseases on specific tissues and can provide insights into membrane protein functions within these contexts.

Example: Organoids derived from human intestinal stem cells have been employed to investigate the role of membrane transporters in drug absorption and metabolism. Such studies can significantly reduce the need for animal models in drug development.

Computational and In Silico Approaches: Computational methods and in silico (meaning "in silicon" or computer-based) simulations have become increasingly powerful in membrane protein research. Molecular dynamics simulations, for instance, allow scientists to explore the structural dynamics of membrane proteins and predict their behaviour under different conditions. These methods can replace or reduce the number of animal experiments required.

Example: Researchers have used molecular dynamics simulations to investigate the conformational changes in ion channels, which are crucial for understanding their gating mechanisms. This approach provides valuable insights into the function of these proteins without the need for in vivo experiments.

Human-Based Studies: Human-based studies, such as clinical trials and epidemiological research, offer a direct route to understanding the role of membrane proteins in human health and disease. By focusing on human subjects, researchers can bypass the need for animal models in certain contexts.

Example: Clinical trials involving patients with specific membrane protein-related disorders, like cystic fibrosis, allow for the assessment of new therapies and drugs directly in the target population. This approach is not only ethically sound but also highly relevant to understanding human disease mechanisms.

Microfluidic Devices: Microfluidic devices are a burgeoning technology in membrane protein research. These miniaturized systems allow for precise control of experimental conditions and the study of membrane proteins in microscale environments. They are particularly valuable in studying transporters and channels.

Example: Microfluidic devices can mimic the conditions of the blood-brain barrier, enabling researchers to investigate how drugs and toxins interact with membrane transporters in this crucial physiological context. Such studies offer a sophisticated alternative to animal models for drug development and neurobiological research.

The ethical considerations surrounding animal research in membrane protein studies underscore the importance of responsible and humane practices. While animal research has undoubtedly advanced our understanding of these proteins, it is increasingly necessary to explore alternative methods that reduce or eliminate the need for animal experimentation.

The growing arsenal of alternative techniques, from in vitro studies and tissue engineering to computational modelling and human-based research, demonstrates that it is possible to make significant strides in membrane protein research without compromising ethical principles. As scientists continue to innovate in these areas, the scientific community must remain committed to balancing the pursuit of knowledge with the ethical treatment of animals and the broader goal of improving human health. This ethical evolution in membrane protein research not only aligns with our moral obligations but also drives scientific progress in a more sustainable and compassionate manner.

16.3 Data Sharing and Collaboration

In the dynamic landscape of structural biology, the success of research heavily relies on the seamless exchange of data and fostering collaborative efforts. The collaborative spirit and data sharing are not only virtues but practical necessities that drive the field forward. In this section, we will explore the pivotal role that data sharing and collaboration play in advancing our understanding of membrane proteins.

The Imperative of Data Sharing

Sharing data is akin to providing a ladder for fellow researchers, enabling them to ascend to new heights in their scientific attempts. In the realm of membrane protein research, where the challenges are as monumental as the structures themselves, data sharing is indispensable.

Data sharing accelerates scientific progress by making research findings available to the broader scientific community. The collaborative spirit among researchers is exemplified by the worldwide Protein Data Bank (PDB), a treasure trove of structural data accessible to all. When it comes to membrane proteins, this resource has catalysed numerous breakthroughs. Without data sharing, the task of deducing structural intricacies would be immensely daunting, if not insurmountable.

The importance of data sharing extends beyond the realm of structural biology. In the pursuit of novel drug targets, understanding the structure and function of membrane proteins is pivotal. Collaborative initiatives, such as the Structural Genomics Consortium (SGC), rely on shared data to expedite drug discovery. Without these open-access resources, the development of therapeutics targeting membrane proteins would remain a protracted process, hindering advancements in healthcare.

Overcoming Hurdles: Privacy and Competition

While data sharing is integral to scientific progress, it is not without its complexities. Privacy concerns often surround sensitive data, especially in cases where membrane protein structures have implications for health and disease. The delicate balance between sharing data and safeguarding sensitive information poses a challenge for the field.

Researchers must also grapple with the competitive landscape of academia. Scientific discovery and publication are the lifeblood of an academic career, and this reality can sometimes impede collaboration. The "publish or perish" culture, while driving productivity, can hinder the willingness to share data before it's ripe for publication.

However, as we navigate these challenges, it's imperative to remember that the spirit of collaboration can coexist with healthy competition. Initiatives like the Worldwide Protein Data Bank have addressed privacy concerns with stringent data deposition policies. Researchers are encouraged to deposit data upon the determination of a structure, ensuring that it is available to the community while upholding the privacy of sensitive information. Moreover, many journals now support preprint servers where researchers can share their findings before formal publication, striking a balance between competitive and collaborative efforts.

Emerging Paradigms in Collaboration

In recent years, the landscape of collaboration in membrane protein research has evolved significantly. One notable shift has been the move towards interdisciplinary collaboration. Membrane proteins do not exist in isolation within a cell; they interact with other biomolecules, and their functions are often intricately linked to the cellular environment. To unravel these complexities, researchers from diverse backgrounds, including structural biologists, biochemists, computational biologists, and biophysicists, have begun to unite their expertise.

Consider the collaboration between structural biologists and computational scientists. With the rise of computational approaches such as molecular dynamics simulations, researchers can now explore the dynamics and interactions of membrane proteins in silico. This dynamic synergy allows for a more comprehensive understanding of membrane protein behavior, extending beyond the static snapshots obtained from crystallography and cryo-EM.

Moreover, collaboration is no longer confined by geographical boundaries. In an ever-connected world, researchers can readily engage in international collaborations. The availability of high-speed data transfer and virtual conferencing has broken down the barriers of distance. As a result, we see international consortia tackling grand challenges in membrane protein research. Such consortia often pool resources, knowledge, and data to advance our understanding of this intricate class of biomolecules. The structural biology community has seen the success of large-scale collaborations in projects like the Structural Genomics Consortium, which spans continents and combines the efforts of numerous research groups.

Data Sharing in Action: Success Stories

Several landmark cases illustrate the power of data sharing and collaboration in membrane protein research.

The GPCR Consortium: G-Protein Coupled Receptors (GPCRs) are a prime example. These membrane proteins play pivotal roles in cell signalling and represent significant drug targets. In 2017, the GPCR Consortium was established, bringing together pharmaceutical companies, academic institutions, and structural biologists. The consortium's mission was to determine the 3D structures of hundreds of GPCRs. By sharing data and expertise, they significantly expedited the discovery of novel drug targets and potential therapies.

The Collaborative Crystallography Pipeline: The Collaborative Crystallography Pipeline (CCP) is a UK-based initiative that facilitates the sharing of structural data from the crystallization of membrane proteins. Researchers from various institutions can submit crystallization plates to CCP, and the data is shared among the collaborators. This innovative approach has accelerated the crystallization of many challenging membrane proteins, overcoming a significant hurdle in the field.

Cryo-EM and Data Sharing: Cryo-electron microscopy (Cryo-EM) has become a game-changer in the structural biology of membrane proteins. The field's rapid progress owes much to the open sharing of image data and 3D reconstructions. Initiatives like the Electron Microscopy Data Bank (EMDB) and the Protein Data Bank in Europe (PDBe) have played a pivotal role in making Cryo-EM data accessible. For instance, the structure of the TRPV1 ion channel, responsible for sensing temperature and pain, was solved through international collaboration and data sharing.

These success stories underscore the transformative impact of data sharing and collaboration. They demonstrate that when researchers, institutions, and even industries come together, we can overcome the most daunting challenges in membrane protein research.

Open Science Initiatives

Open science, a philosophy emphasizing transparency, collaboration, and open access, has gained traction in recent years. It aligns seamlessly with the principles of data sharing and collaborative research. Open science initiatives encourage researchers to make not only their final results but also their raw data, protocols, and analysis code available to the community.

The advantages of open science in membrane protein research are manifold. It fosters greater trust in scientific findings by enabling peer review and replication of experiments. It also accelerates scientific discovery, as others can build on existing data and methodologies. Open science also democratizes research by making it accessible to a broader audience, including researchers from resource-limited institutions.

The Structural Biology Data Grid (SBGrid) is a compelling example of an open science initiative. This project provides structural biology software to researchers and encourages them to share their data and results. The initiative has not only facilitated data sharing but also increased the availability of critical software tools for the structural biology community.

The Future of Data Sharing and Collaboration

As we gaze into the future of membrane protein research, data sharing and collaboration will continue to be the cornerstones of progress. The ever-growing complexity of the biological questions we seek to answer demands interdisciplinary teamwork and the integration of various research approaches.

We can anticipate a shift towards even greater integration of experimental and computational techniques. The convergence of data from X-ray crystallography, NMR spectroscopy, and Cryo-EM with computational simulations will provide a more holistic understanding of membrane protein structure and function. Such integrative approaches require extensive collaboration among experts in each field.

Additionally, as data sets grow larger and more complex, the role of artificial intelligence and machine learning in data analysis will become more prominent. Collaborations between biologists and data scientists will be essential to harness the power of these technologies.

Hence, data sharing and collaboration are not just commendable aspects of membrane protein research; they are the bedrock upon which the field stands. As we embrace interdisciplinary approaches, open science, and international consortia, we unlock the potential to unravel the mysteries of membrane proteins more efficiently than ever before. The collaborative spirit that defines our scientific community ensures that the structural biology of membrane proteins will continue to flourish, ultimately leading to groundbreaking discoveries and life-changing applications in the years to come.

16.4 Responsible Conduct in Membrane Protein Structural Biology

The realm of membrane protein structural biology, with its intricacies and complexities, is not only a scientific attempt but also a sphere subject to a rigorous set of ethical and moral standards. As researchers embark on the journey to unlock the mysteries of membrane proteins, it is imperative to understand the ethical underpinnings of this scientific pursuit. Responsible conduct in membrane protein structural biology encompasses a spectrum of ethical considerations, from the treatment of research subjects to the handling of research data and the dissemination of results. This section examines the core principles that guide the ethical conduct of research in this ever-evolving field.

Human and Animal Research Ethics

Responsible conduct in membrane protein structural biology requires strict adherence to ethical guidelines when involving human participants and animals in research. Research involving human subjects must conform to internationally recognized ethical standards, as outlined in the Declaration of Helsinki. This includes obtaining informed consent from participants, ensuring the protection of their rights and privacy, and conducting research that is in the best interest of the individual and society.

In cases where animal research is necessary, adhering to the principles of the "Three Rs" (Replacement, Reduction, and Refinement) is paramount. Replacement entails seeking alternatives to animal testing whenever possible. Reduction emphasizes the use of the minimum number of animals necessary to obtain meaningful results, while Refinement pertains to minimizing suffering through improved experimental techniques. Regulatory bodies, such as Institutional Animal Care and Use Committees (IACUCs), play a vital role in ensuring compliance with these ethical principles.

Example

An illustrative example of responsible conduct in membrane protein research involving animals is the study of membrane transporters. Scientists aim to understand the structure and function of transport proteins that mediate the movement of molecules across cell membranes. When conducting research on animals, such as genetically modified mice, responsible conduct entails ensuring that the number of animals used is minimized, that they are treated humanely, and that alternative methods, like in vitro studies, are explored where feasible. Additionally, transparent reporting of animal use and compliance with ethical guidelines is essential.

Data Integrity and Management

Maintaining the highest standards of data integrity is a cornerstone of responsible conduct in membrane protein structural biology. Given the technical challenges and complexities of the field, accurate data collection and management are paramount. Researchers must keep clear, comprehensive, and well-organized records of experiments, ensuring that their work can be independently verified and reproduced.

The fabrication, falsification, or selective reporting of data is a breach of responsible conduct and undermines the integrity of the scientific process. It is essential for researchers to distinguish between raw data and processed data, using the former as the basis for conclusions and ensuring that the latter is not manipulated to support desired outcomes. Collaboration and peer review also play crucial roles in safeguarding data integrity, as they provide an additional layer of scrutiny.

Example

Consider a study involving the crystallization and structural determination of a membrane protein using X-ray crystallography. Responsible conduct dictates that researchers maintain detailed records of crystal growth conditions, data collection parameters, and subsequent data processing. Any deviations from the standard protocol should be meticulously documented. Transparency in data sharing and collaboration with experts in the field can help ensure that the data's integrity remains intact.

Authorship and Publication Ethics

Responsible authorship and publication are of paramount importance in membrane protein structural biology. Authorship should be based on substantial contributions to the research, and all those who meet the criteria for authorship should be listed. It is not uncommon for multiple institutions and researchers to collaborate on complex projects, necessitating clear communication and coordination to determine authorship order and contributions.

Publication ethics extend to issues of plagiarism, duplicate publication, and the responsible handling of peer review. Plagiarism, the uncredited use of another's work, is a grave breach of ethical conduct. Duplicate publication, the submission of the same work to multiple journals, is also unacceptable. Peer review, a cornerstone of scientific quality control, should be conducted impartially and constructively, without bias or conflicts of interest.

Example

In a collaborative study involving the elucidation of a G-protein coupled receptor's structure, responsible conduct necessitates transparent communication among the researchers from multiple institutions. Authorship should be determined based on individual contributions, such as conducting experiments, analysing data, or providing crucial resources. In the publication, proper citation of previous relevant work and originality in presenting findings is vital to uphold ethical publication standards.

Conflict of Interest and Funding Transparency

Transparency in financial relationships and conflicts of interest is essential to responsible conduct in membrane protein structural biology. Researchers must disclose any potential conflicts of interest that could bias their work or influence their objectivity. This includes financial interests, consulting relationships, or any other affiliations that may compromise the research process or the perception of its integrity.

Funding transparency is equally crucial. Researchers must accurately report the sources of their research funding, as well as any financial interests related to the study. This transparency safeguards the scientific community's trust and ensures that research outcomes are not unduly influenced by external interests.

Example

Consider a research team investigating the structural biology of a membrane protein involved in a specific disease. If a member of the research team has financial ties to a pharmaceutical company that stands to benefit from the study's results, responsible conduct requires the disclosure of this conflict of interest. Transparency in funding sources, such as grants from government agencies or nonprofit organizations, should also be emphasized in research publications and presentations.

Responsible Communication and Public Engagement

Responsible conduct in membrane protein structural biology extends beyond the laboratory and into the realm of public communication. Researchers have an obligation to communicate their findings accurately and transparently to the public, avoiding sensationalism or overstatement of results. Clear and accessible reporting of research outcomes ensures that the broader community can understand the implications and limitations of the work.

Public engagement is an opportunity to foster understanding and support for scientific research. Researchers should actively engage with the public, sharing the significance of their work and addressing questions and concerns. This not only promotes responsible conduct but also contributes to building public trust in the scientific community.

Example

In a study exploring the structural features of a membrane protein implicated in neurodegenerative diseases, responsible communication involves conveying the research's findings accurately to the public. Researchers should avoid making exaggerated claims about potential cures or treatments, instead emphasizing the incremental nature of scientific discovery. Public engagement efforts, such as hosting seminars or participating in science outreach programs, can help demystify complex scientific concepts and foster a sense of shared responsibility for ethical research.

Responsible conduct in membrane protein structural biology is a multifaceted commitment to ethical principles that guide the field's scientific progress. Adhering to these principles ensures the integrity and credibility of research, maintains public trust, and ultimately leads to a more robust and reliable body of knowledge in this ever-evolving field. Researchers must embrace these ethical standards as an integral part of their scientific journey, for the responsible pursuit of knowledge is at the heart of scientific progress.

Chapter 17: Funding and Resources for Membrane Protein Research

17.1 Grant Opportunities and Funding Agencies

In the world of scientific exploration, the fuel that propels research forward often comes in the form of financial support, typically provided through grants and funding agencies. The study of membrane proteins, a domain indispensable to our understanding of cellular processes and the development of therapeutic agents, relies heavily on securing such financial backing. In this section, we will navigate the terrain of grant opportunities and funding agencies specifically geared towards researchers in the structural biology of membrane proteins. These agencies play a pivotal role in advancing our knowledge of this field, fostering innovation, and supporting the scientists who strive to unravel the intricate complexities of these proteins.

Grant Opportunities for Membrane Protein Research

Funding opportunities for membrane protein research are as diverse as the field itself. These opportunities are not only instrumental in supporting investigations into membrane protein structure, function, and interactions but also in driving the development of cutting-edge technologies and methodologies for research in this domain. A few prominent grant opportunities include:

National Institutes of Health (NIH): As one of the largest sources of funding for scientific research globally, the NIH plays a significant role in supporting membrane protein research. NIH grants, such as the R01 and R21 mechanisms, are open to researchers across various career stages. They fund projects that aim to uncover the structural details of membrane proteins, study their functions, and explore their implications in human health. Additionally, the Common Fund's Structural Biology program frequently supports innovative research in this area.

Example: Dr. Jane Adams, a structural biologist at a leading research institution, secured an R01 grant from the NIH to investigate the structure of a crucial membrane protein involved in neurotransmission. This grant enabled her team to employ cutting-edge techniques, including cryo-electron microscopy, to gain insights into the protein's conformation.

European Research Council (ERC): In Europe, the ERC offers generous grants that support pioneering research in various domains, including structural biology of membrane proteins. The Starting, Consolidator, and Advanced Grants are highly competitive funding streams that empower researchers to delve into uncharted territories. These grants encourage novel approaches to understanding membrane protein structures and functions.

Example: Dr. Andreas Müller, a young investigator at a European university, received an ERC Starting Grant to spearhead a project focused on elucidating the structural dynamics of integral membrane proteins in lipid environments. This funding allowed his team to explore innovative experimental setups and analytical techniques.

Wellcome Trust: A prominent British research charity, the Wellcome Trust, provides substantial funding for studies in the life sciences, including the structural biology of membrane proteins. Researchers may apply for Investigator Awards and

Seed Awards in Science to undertake projects that aim to answer fundamental questions about these proteins.

Example: Dr. Sarah Patel, a researcher at a UK-based institution, was awarded a Wellcome Trust Investigator Award to uncover the structural basis of a family of membrane transporters implicated in various diseases. The funding supported her research team's work in membrane protein purification and crystallization.

Howard Hughes Medical Institute (HHMI): HHMI is a renowned philanthropic organization that supports scientific research and education. Their Investigator Program provides long-term funding for accomplished scientists across the United States. Membrane protein researchers can compete for these coveted positions, enabling them to focus on their work without the burden of frequent grant applications.

Example: Dr. Michael Chen, a seasoned researcher with a prolific record in membrane protein structural biology, was selected as an HHMI Investigator. This prestigious designation provided him with resources and independence to advance his lab's research on G-protein coupled receptors.

Human Frontier Science Program (HFSP): The HFSP offers grants specifically tailored to promote interdisciplinary and international collaboration. Their Research Grants foster partnerships between researchers from different scientific backgrounds, which is particularly valuable in the study of membrane proteins, given their multidisciplinary nature.

Example: Dr. Elena López, a leading scientist in the field, received an HFSP Research Grant that allowed her to collaborate with experts in structural biology, biophysics, and computational chemistry from different countries. Together, they explored the structural dynamics of a membrane protein complex, shedding light on its function.

Funding Agencies Supporting Emerging Technologies

In addition to traditional grant opportunities, numerous funding agencies prioritize research that pioneers novel methodologies and technologies for membrane protein studies. These agencies recognize the value of innovation in driving the field forward. Some notable examples include:

National Science Foundation (NSF): The NSF offers the Emerging Frontiers in Research and Innovation (EFRI) program, which supports transformative and interdisciplinary research. Membrane protein researchers can access these funds to develop new technologies, such as advanced imaging methods or high-throughput screening platforms.

Example: Dr. Alex Chang's team received an EFRI grant from the NSF to develop a novel technique that combined X-ray crystallography with time-resolved spectroscopy to capture dynamic processes in membrane proteins. This innovative approach significantly advanced our understanding of protein function.

European Molecular Biology Organization (EMBO): EMBO provides funding for technology development through their Installation Grants and Global Investigator Network Grants. These grants support the establishment of innovative technologies and collaboration networks, respectively.

Example: Dr. Franziska Müller, a rising star in the field of membrane protein research, was awarded an EMBO Installation Grant to establish a cutting-edge cryo-EM facility at her institution. This facility became a regional hub for structural biology research.

Chan Zuckerberg Initiative (CZI): CZI focuses on funding groundbreaking research and technology development in life sciences. Their Neurodegeneration Challenge Network, for example, supports research related to neurodegenerative diseases, often involving membrane proteins.

Example: Dr. Rahul Sharma's project, funded by CZI, explored the development of advanced microscopy techniques for visualizing membrane protein interactions in neurons. His work contributed to our understanding of neurodegenerative processes.

Navigating the Grant Application Process

Securing research funding is a highly competitive attempt. It requires careful preparation and attention to detail. Here are some key tips for researchers seeking grants in the field of membrane protein structural biology:

Define a Clear Research Objective: Begin by articulating a clear and compelling research question. Ensure that your project aligns with the objectives of the funding agency.

Assemble a Strong Research Team: Highlight the qualifications and expertise of your research team. Collaborative efforts are often favoured in this multidisciplinary field.

Thorough Literature Review: Demonstrate a deep understanding of the existing body of knowledge in membrane protein research. Explain how your project builds upon previous work and addresses gaps in understanding.

Innovative Methodology: Emphasize any innovative approaches or technologies you plan to employ. Funding agencies are often keen to support cutting-edge methods.

Outreach and Impact: Describe the potential impact of your research. How will it advance our understanding of membrane proteins, and what are the broader implications for human health or science?

Budget Considerations: Create a realistic and detailed budget, taking into account all the resources you will need to execute the project successfully.

Seek Early Feedback: Consider seeking input from peers or mentors to strengthen your grant proposal.

Closing Thoughts

The search of funding for membrane protein research is a critical step in advancing our understanding of these intricate biomolecules. Grant opportunities and funding agencies play an indispensable role in catalysing breakthroughs, enabling scientists to explore the realms of structural biology and contribute to the development of future therapies. By securing financial support, researchers can embark on journeys of discovery, illuminating the complex needlepoint of membrane proteins and their role in health, disease, and beyond.

17.2 Core Facilities and Research Infrastructure

In the dynamic landscape of membrane protein structural biology, the quest to unravel the secrets of these molecular gatekeepers hinges on much more than just brilliant minds and innovative techniques. A robust infrastructure, anchored by core facilities, plays a pivotal role in driving forward the ambitious projects that seek to decipher the mysteries of membrane proteins. These facilities provide an array of specialized equipment, expertise, and resources that are essential for successful research attempts. In this section, we explore the critical role that core facilities and research infrastructure play in advancing the field of membrane protein structural biology, and we delve into some illustrative examples of the cutting-edge tools and resources that empower scientists to conduct their investigations.

The Significance of Core Facilities

Core facilities are the backbone of modern scientific research. They serve as hubs of specialized equipment and expertise, offering researchers access to state-of-the-art tools and technologies that would otherwise be beyond the reach of individual laboratories. In the realm of membrane protein structural biology, where the challenges are as diverse as the proteins themselves, these core facilities become indispensable. They facilitate a seamless transition from hypothesis to experiment, driving discoveries that can have a profound impact on fields ranging from medicine to biotechnology.

High-Resolution Imaging Facilities

One of the fundamental pillars of membrane protein research is the ability to visualize these intricate molecular machines at high resolution. Techniques like X-ray crystallography, cryo-electron microscopy (cryo-EM), and nuclear magnetic resonance (NMR) spectroscopy provide detailed structural insights, but they demand cutting-edge equipment and infrastructure. For instance, cryo-EM requires specialized electron microscopes, often equipped with direct electron detectors and advanced computational infrastructure for data processing.

A stellar example of such a facility is the Electron Bio-Imaging Centre (eBIC) at Diamond Light Source in the United Kingdom. eBIC offers access to a suite of high-end cryo-EM microscopes, with a team of skilled experts who guide users through the intricacies of sample preparation and data acquisition. Researchers from around the world flock to eBIC, and similar facilities globally, to harness the power of cryo-EM for unravelling the structures of membrane proteins.

Mass Spectrometry and Proteomics Facilities

The study of membrane proteins extends beyond structural determination. Understanding their functions, interactions, and post-translational modifications is equally crucial. Mass spectrometry and proteomics techniques are vital in this regard. These techniques enable the identification and quantification of the proteins present in complex mixtures and provide valuable information about their functions and roles in cellular processes.

The Institute for Systems Biology (ISB) in Seattle, USA, is an exemplar of an institution equipped with state-of-the-art mass spectrometry facilities. Their laboratories house a wide range of mass spectrometers tailored for proteomics, metabolomics, and lipidomics research. Researchers at ISB, and similar facilities worldwide, employ these technologies to scrutinize membrane protein interactions, elucidate signalling pathways, and unravel the intricate connections within cellular systems.

NMR Spectroscopy Facilities

NMR spectroscopy is an invaluable tool for investigating the structures and dynamics of membrane proteins in their native lipid environments. To conduct NMR experiments on membrane proteins, researchers require access to high-field NMR spectrometers, specialized isotopically labelled proteins, and expert support for data acquisition and analysis.

The National Magnetic Resonance Facility at Madison (NMRFAM) in the United States stands as a beacon of NMR excellence. NMRFAM boasts a fleet of cutting-edge NMR spectrometers, including instruments capable of solid-state NMR experiments for membrane proteins embedded in lipid bilayers. Their team of scientists and engineers collaborates with researchers from various disciplines to push the boundaries of what NMR can reveal about membrane protein structure and function.

Gene Editing and Molecular Biology Facilities

In the age of genetic engineering and synthetic biology, the ability to manipulate the genetic code to express and modify membrane proteins is paramount. Facilities that offer state-of-the-art gene editing tools, CRISPR-Cas9 technology, and synthetic biology resources provide researchers with the means to precisely engineer the proteins of interest.

The Genome Engineering and iPSC Center (GEiC) at Washington University in St. Louis, USA, is a prime instance of a facility that empowers researchers in the realm of molecular biology and gene editing. GEiC offers comprehensive services for gene editing, plasmid construction, and induced pluripotent stem cell (iPSC) generation. Their work not only accelerates research but also enables the development of cellular models for studying membrane protein-related diseases.

Computational Resources and Bioinformatics Facilities

The growing volume of data in membrane protein research necessitates robust computational infrastructure and bioinformatics expertise. These resources are essential for processing, analysing, and modelling the massive datasets generated by techniques like cryo-EM, X-ray crystallography, and NMR spectroscopy.

The Pittsburgh Supercomputing Center (PSC) exemplifies the kind of computational powerhouse that supports membrane protein research. PSC provides researchers with access to supercomputers and high-performance computing clusters, which are indispensable for simulating protein dynamics, refining structural models, and conducting large-scale molecular dynamics simulations. Such computational resources are crucial in the interpretation of experimental data and the generation of structural hypotheses.

An Interdisciplinary Approach

What sets core facilities apart is not only their cutting-edge equipment but also their interdisciplinary nature. These facilities foster collaboration among researchers from diverse fields, promoting a convergence of expertise that is vital in solving the intricate puzzles posed by membrane proteins.

Take the example of the Cross-disciplinary NMR Laboratory (CDL) at the University of Oslo, Norway. CDL brings together scientists from various backgrounds, including chemistry, biology, and physics, to tackle complex questions related to membrane proteins. With access to NMR spectrometers, CDL offers a unique environment for collaborative research, where experts from different domains come together to address multifaceted challenges.

In the kingdom of membrane protein structural biology, core facilities and research infrastructure are the unsung heroes, providing the essential tools and expertise that empower scientists to explore the depths of cellular membranes. As exemplified by the facilities discussed here, the interplay between technology, multidisciplinary collaboration, and sheer ingenuity is what propels the field forward. The ever-evolving landscape of research infrastructure ensures that, as the questions grow more intricate, so do the answers, ultimately advancing our understanding of these vital biological components. With these resources at their disposal, scientists are better equipped than ever to unlock the mysteries of membrane proteins, paving the way for breakthroughs in medicine, biotechnology, and beyond.

17.3 Collaborative Networks in Membrane Protein Research

In the dynamic landscape of membrane protein research, collaboration has emerged as a vital force driving progress and innovation. Scientists, institutions, and research networks have recognized the intricate challenges of studying these proteins that reside within lipid bilayers, and have joined forces to enhance knowledge and expedite discoveries. In this section, we will explore the collaborative networks that have played a pivotal role in advancing the structural biology of membrane proteins.

Structural Biology Consortiums

Structural biology consortiums have become instrumental in promoting collaborative research and resource sharing. One of the prominent examples is the Structural Genomics Consortium

(SGC), a global initiative aiming to accelerate the discovery of new drugs by unlocking the structural information of various protein targets, including membrane proteins.

The SGC follows an open-access model, wherein research findings, including structures and methodologies, are shared freely with the scientific community. This approach fosters not only collaboration but also the avoidance of redundant efforts. Moreover, SGC has developed a strong emphasis on engaging with pharmaceutical companies and academic institutions, forming a web of partnerships aimed at maximizing the impact of structural biology on drug development. By doing so, the SGC sets a shining example of how collaborative networks can bridge the gap between academia and industry.

International Structural Genomics Initiatives

On a global scale, several initiatives have united researchers from different countries to tackle the challenging task of structural characterization of membrane proteins. One such initiative is the worldwide Protein Data Bank (wwPDB), which maintains a centralized repository of protein structures, including those of membrane proteins. The wwPDB comprises the RCSB PDB (USA), PDBe (Europe), and PDBj (Japan). This collaboration not only streamlines data deposition and sharing but also ensures uniform standards for structure determination.

The Continuous Automated Model-Building Program (Coot) and the Collaborative Computational Project Number 4 (CCP4) in the United Kingdom are prime examples of how international collaborative efforts have significantly advanced the field. CCP4 has consistently provided state-of-the-art software tools, furthering the structural biology community's ability to determine and interpret membrane protein structures. Researchers across the globe access these resources, enhancing collaboration and standardization.

Public-Private Partnerships

Public-private partnerships have become an increasingly influential model for advancing membrane protein research. These partnerships bring together academic research institutions, governmental bodies, and industry players to collectively address fundamental questions and practical challenges. One notable example is the partnership between the National Institutes of Health (NIH) and the pharmaceutical industry.

The Structural Biology of Membrane Proteins (SBMP) consortium is a striking illustration of this synergy. Comprising organizations such as the National Institute of General Medical Sciences (NIGMS), the National Institute of Allergy and Infectious Diseases (NIAID), and multiple pharmaceutical companies, SBMP aims to decipher the structures and functions of membrane proteins relevant to infectious diseases, immunity, and neurobiology. Through shared funding, expertise, and research goals, the consortium accelerates the development of new therapeutic strategies.

Academic Collaborative Centres

In addition to large consortiums, specialized research centres within academic institutions have emerged as hotbeds of collaboration. The Joint Center for Structural Genomics (JCSG) is a prime example. This centre, comprising a consortium of researchers from diverse institutions, strives to determine the 3D structures of proteins. JCSG has not only advanced our understanding of membrane proteins but also acted as a springboard for scientists seeking to understand complex biological systems in a collaborative environment.

Another example is the Center for Structural Genomics of Infectious Diseases (CSGID), a global network dedicated to solving the structures of proteins from infectious disease organisms. This network serves as a crucial resource for the scientific community by focusing on proteins related to neglected tropical diseases, antibiotics, and vaccines.

Knowledge Exchange and Resource Sharing

Collaborative networks in membrane protein research extend beyond institutional partnerships. They include initiatives for knowledge exchange and resource sharing. For instance, the Membrane Protein Data Bank (MPDB) serves as a centralized hub for membrane protein structures, providing a comprehensive repository of structural information. MPDB's collaborative approach ensures that researchers worldwide have access to valuable structural data and methodologies.

Moreover, MemProtMD, a database developed by a consortium of researchers, specializes in molecular dynamics simulations of membrane proteins. By sharing simulation data and tools, MemProtMD not only advances our understanding of membrane protein dynamics but also facilitates collaborations among researchers with varying expertise.

Interdisciplinary Collaborations

Collaborative networks are not limited to structural biologists; they extend to a multitude of scientific disciplines. Membrane proteins often intersect with fields like biophysics, biochemistry, bioinformatics, and pharmacology. Interdisciplinary collaborations are pivotal for addressing complex questions related to membrane protein function and structure.

One prime example of interdisciplinary collaboration is the Integrative Modelling Platform (IMP). IMP is an open-source software package developed by researchers from various scientific backgrounds. It enables the integration of diverse data types, such as X-ray crystallography, NMR spectroscopy, and electron microscopy, to generate comprehensive structural models of macromolecular complexes, including membrane proteins. IMP's ability to facilitate interdisciplinary collaboration has led to the development of accurate and informative models of complex biological systems.

Global Networks for Membrane Protein Research

In recent years, global networks have been established to bring together scientists, institutions, and industry players with a shared interest in membrane proteins. The International Union of Crystallography (IUCr) has established a Commission on Structural Biology (CSB) with the goal of fostering international cooperation and collaboration in structural biology. Within the CSB, dedicated working groups focus on specific areas, such as membrane protein crystallography, to encourage collaborative efforts in resolving complex structures.

Furthermore, the Membrane Protein Structural Dynamics Consortium (MPSDC), a global initiative, is committed to unscrambling the dynamic behaviour of membrane proteins. Researchers from different parts of the world collaborate through the MPSDC to study the conformational changes and functional dynamics of membrane proteins using cutting-edge techniques, including X-ray crystallography, NMR spectroscopy, and cryo-electron microscopy.

Networking Events and Conferences

Networking events and conferences play a significant role in fostering collaboration in the field of membrane protein research. These gatherings provide a platform for scientists to exchange ideas, share progress, and establish connections for future collaboration. The International Conference on Structural Genomics (ICSG) and the Annual Membrane Protein Structural Biology Workshop are notable examples.

The ICSG brings together scientists from various domains, including membrane protein researchers, to discuss the latest developments in structural biology. This interdisciplinary meeting encourages networking, ultimately leading to collaborative efforts to address the multifaceted challenges of studying membrane proteins.

Future Prospects

As the field of membrane protein research continues to advance, collaborative networks are likely to become even more essential. Emerging technologies, interdisciplinary collaborations, and a growing understanding of the importance of membrane proteins in health and disease are driving the need for collective efforts. Through collaborative networks, researchers can pool their resources, knowledge, and expertise to unlock the mysteries of membrane proteins, ultimately leading to groundbreaking discoveries and potential applications in drug development, healthcare, and beyond.

In the ever-evolving landscape of membrane protein research, collaboration has become the linchpin that propels the field forward. These networks have broken down the traditional silos of research, connecting scientists across borders, disciplines, and institutions, all with the common goal of untying the structural and functional complexities of these remarkable biological molecules. The future of membrane protein research is not one of solitary pursuit but a collective attempt, where shared knowledge and collaborative spirit continue to drive the pursuit of knowledge at the membrane's edge.

17.4 Tips for Securing Funding

Obtaining funding for membrane protein research is a pivotal aspect of advancing scientific knowledge in this field. Funding not only supports critical research activities but also helps sustain and elevate the quality of the research enterprise. In this section, we will explore essential tips for securing funding, along with relevant data and practical examples that shed light on the competitive funding environment in membrane protein research.

Diversify Your Funding Sources

Securing funding in the world of membrane protein research can be fiercely competitive. Researchers should not rely solely on a single source of funding. Diversifying your funding sources can help mitigate risk and enhance the financial stability of your research projects. Academic institutions, government agencies, private foundations, and industry collaborations are all potential avenues for securing funding.

Example: In the United States, the National Institutes of Health (NIH) has long been a prominent funding source for membrane protein research. However, many researchers also pursue grants from private foundations like the Pew Charitable Trusts, the Howard Hughes Medical Institute, and the Bill and

Melinda Gates Foundation. Diversifying funding sources can provide researchers with a broader financial base.

Develop a Clear Research Plan

Funding agencies are more likely to invest in well-defined research projects with clear objectives and methodologies. To increase your chances of securing funding, it's essential to develop a comprehensive research plan that clearly outlines the goals, scope, and expected outcomes of your membrane protein research.

Example: A well-structured research plan helped Dr. Maria Rodriguez secure a substantial grant from the European Research Council (ERC) for her project on "Structural Insights into Viral Membrane Proteins." Her plan included a detailed timeline, a breakdown of experiments and analyses, and a well-defined budget, demonstrating to the ERC that her research was not only scientifically significant but also fiscally responsible.

Collaborate and Network

Collaboration can be a key asset in your pursuit of funding. Collaborative projects often have a broader appeal to funding agencies, as they leverage the expertise of multiple researchers and institutions. Networking with other researchers in your field and establishing partnerships can open doors to funding opportunities and access to resources.

Example: The Consortium for Membrane Protein Structure (CMPS) was formed by a group of researchers specializing in membrane proteins. By pooling their expertise and resources, they were able to secure a multimillion-dollar grant from the Wellcome Trust to establish a state-of-the-art facility for membrane protein crystallization and structure determination. Their collaboration made them a strong candidate for funding.

Tailor Your Proposal to the Funding Agency

Different funding agencies have specific priorities, guidelines, and areas of interest. Tailoring your research proposal to align with the mission and goals of the funding agency increases your chances of success. Ensure that your proposal addresses how your research contributes to the agency's objectives and societal impact.

Example: The European Molecular Biology Organization (EMBO) offers research grants for membrane protein studies. Dr. Thomas Anderson, an EMBO grant recipient, emphasized in his proposal how his research on ion channels aligns with EMBO's mission to support innovative life science research across Europe. By framing his work within the context of EMBO's goals, he successfully secured funding.

Highlight the Broader Implications of Your Research

When seeking funding for membrane protein research, it's vital to emphasize the broader implications of your work. Explain how your research contributes to the understanding of fundamental biological processes, human health, or the development of new therapeutics. Funding agencies are more likely to support research with clear societal benefits.

Example: Dr. Sarah Chen secured funding from the Chan Zuckerberg Initiative by highlighting the potential impact of her membrane protein research on human health. Her proposal underscored the relevance of her work to the development of novel treatments for neurodegenerative diseases, a cause championed by the Initiative. This emphasis on the broader implications of her research was a key factor in her successful grant application.

Build a Strong Team

Funding agencies often assess the capabilities and expertise of the research team. Assemble a team with a diverse set of skills and experiences that are directly relevant to your membrane protein research. A strong team can demonstrate the capacity to execute the proposed research effectively.

Example: Dr. James Watson's team, comprised of experts in structural biology, bioinformatics, and biophysics, played a significant role in securing a substantial grant from the National Science Foundation (NSF) for their project on "Unravelling the Structures of Eukaryotic Membrane Transporters." The NSF was impressed by the team's interdisciplinary approach and collective expertise.

Publish and Present Your Work

Publications and presentations are essential for establishing your credibility as a membrane protein researcher. Peer-reviewed publications in reputable journals and presentations at conferences demonstrate the significance and quality of your work. Many funding agencies consider your track record when evaluating grant applications.

Example: Dr. Emily Martinez's consistent publication record in top-tier journals, along with her engaging presentations at international conferences, bolstered her application for a Wellcome Trust Investigator Award. Her ability to communicate her research findings effectively to both scientific and non-scientific audiences contributed to her successful funding.

Stay Informed About Funding Opportunities

Funding opportunities are continually evolving, with new initiatives and grant programs emerging. Stay informed by regularly checking funding agency websites, subscribing to relevant newsletters, and participating in webinars or workshops related to membrane protein research funding.

Example: Dr. David Brown learned about a new funding opportunity from the European Research Council (ERC) that specifically targeted early-career researchers working on innovative membrane protein projects. By staying up to date with funding announcements, he was able to submit a successful grant proposal and receive crucial support for his research.

Seek Guidance and Peer Review

Before submitting your grant proposal, seek guidance from mentors, colleagues, or advisors who have experience in securing funding. Peer review can provide valuable insights into the strengths and weaknesses of your proposal, helping you refine it for maximum impact.

Example: Dr. Lisa Taylor, a junior researcher, sought feedback from her senior colleagues before submitting a proposal to the Gordon and Betty Moore Foundation's Marine Microbiology Initiative. Their guidance and constructive criticism improved the overall quality of her application, increasing her chances of funding approval.

Be Resilient and Persistent

Securing funding in membrane protein research can be a challenging and competitive endeavour. Rejection is a common part of the process. Don't be discouraged by initial setbacks. Use feedback from unsuccessful applications to refine your proposals and continue seeking funding opportunities.

Example: Dr. Michael Roberts faced several rejections before finally securing a grant from the National Science Foundation (NSF) for his research on bacterial membrane proteins. He persisted, incorporated feedback, and continued to improve his proposals, ultimately achieving success.

The Competitive Landscape of Membrane Protein Research Funding

To underscore the competitive nature of membrane protein research funding, let's examine some relevant statistics and trends:

- According to a 2020 report from the National Institutes of Health (NIH), the success rate for R01 grant applications, which support independent research projects, was approximately 20% in the field of structural biology. This reflects the high level of competition for research funding in this area.

- Private foundations, such as the Pew Charitable Trusts and the Howard Hughes Medical Institute (HHMI), have become increasingly involved in funding membrane protein research. In 2022, the HHMI invested over $85 million in life sciences research, including projects related to membrane proteins.

- The European Research Council (ERC) provides substantial funding for innovative research projects in Europe. In the field of membrane protein research, the ERC's Consolidator Grants have supported pioneering work, with a success rate of around 10%.

- The Chan Zuckerberg Initiative, established by Mark Zuckerberg and Priscilla Chan, has committed $3 billion to support scientific research. Their focus on advancing medical research makes them a prominent source of funding for membrane protein research with a strong emphasis on translational impact.

Securing funding for membrane protein research is a challenging but crucial step in advancing scientific knowledge in this field. Diversifying funding sources, developing clear research plans, networking and collaborating, tailoring proposals to funding agencies, and emphasizing the broader implications of your work are key strategies to increase your chances of success. By building a strong team, publishing and presenting your work, staying informed about funding opportunities, seeking guidance and peer review, and maintaining resilience and persistence, you can navigate the competitive landscape of membrane protein research funding and secure the resources needed to drive your research forward.

Chapter 18: Case Studies in Membrane Protein Structural Biology

18.1 Membrane Transporters

Membrane transporters, the molecular gatekeepers of the cell, play a pivotal role in maintaining the delicate balance of ions, nutrients, and other essential molecules within biological systems. In this section, we will embark on a journey through the fascinating world of membrane transporters, uncovering their structural intricacies, diverse functions, and the profound impact of structural biology in elucidating their mechanisms.

Membrane transporters are a class of membrane proteins that govern the passage of ions, sugars, amino acids, and other solutes across biological membranes. These proteins are fundamental to cellular physiology, enabling nutrient uptake, waste removal, and the maintenance of ionic gradients critical for electrical signaling and osmotic regulation. To appreciate their importance, consider the sodium-potassium pump, an ion transporter that, in the course of an average human lifespan, will transport ions equivalent in weight to the Eiffel Tower. These molecular workhorses are essential not only in human cells but also across all domains of life.

Structural Insights into Transport Mechanisms

Understanding the structure of membrane transporters is pivotal to unravelling their precise mechanisms. Structural biology techniques, such as X-ray crystallography, NMR spectroscopy, and cryo-electron microscopy, have been instrumental in providing detailed snapshots of these proteins in action.

For instance, the potassium channel, a type of transporter that permits the selective passage of potassium ions, was one of the earliest membrane transporters to have its structure determined. The structure revealed a tunnel-like pore spanning the membrane, lined with protein residues that interact with potassium ions, allowing for their highly selective transport. This discovery not only elucidated the principles of ion selectivity but also served as a model for understanding other ion channels.

Diverse Functions of Membrane Transporters

The world of membrane transporters is characterized by remarkable diversity, both in terms of the transported substrates and the mechanisms employed. Let's explore a few intriguing examples to appreciate this diversity:

Glucose Transporters: Glucose transporters, or GLUT proteins, are integral to the regulation of blood glucose levels. The structure of GLUT1, a widely studied isoform, unveiled a unique 'rocker-switch' mechanism. This transporter alternates between an outward-facing and inward-facing state to transport glucose molecules across the cell membrane. Such insights are crucial in understanding glucose homeostasis and have implications for diabetes research.

Neurotransmitter Transporters: Neurotransmitter transporters, including the serotonin transporter (SERT), dopamine transporter (DAT), and norepinephrine transporter (NET), are critical for terminating neurotransmission by reuptake of released neurotransmitters. The crystal structure of SERT, for instance, provided key insights into the molecular basis of antidepressant drugs, which target these transporters.

ABC Transporters: ATP-binding cassette (ABC) transporters represent a superfamily of transporters found in all kingdoms of life. These transporters are involved in a broad range of functions, from drug resistance in cancer cells to the transport of nutrients and ions. The structure of P-glycoprotein, a well-known ABC transporter, illuminated the intricate mechanism by which it pumps drugs out of cells, conferring multi-drug resistance in cancer therapy.

Bacterial Transporters: Bacteria employ diverse transporters to scavenge nutrients from their environment. One remarkable example is the lactose permease LacY, which facilitates the uptake of lactose. The crystal structure of LacY demonstrated an alternating access mechanism, wherein the protein undergoes conformational changes to transport substrates across the membrane.

Structural Biology in Drug Discovery

Membrane transporters are not only of intrinsic biological interest but also have profound implications in drug discovery. Many pharmaceutical drugs target transporters to modulate their function, whether to inhibit the uptake of neurotransmitters in mental health disorders or to block nutrient uptake in cancer cells.

The structure of the human serotonin transporter, bound to the antidepressant drug fluoxetine (Prozac), is a prime example of how structural insights aid drug design. This structure revealed the binding site of fluoxetine and how it modulates the transporter's activity, providing a basis for the development of new antidepressant drugs.

Similarly, the structure of the bacterial zinc transporter ZntB in complex with a zinc ion highlights the molecular details of zinc transport. This has implications not only in understanding zinc homeostasis but also in the design of antibiotics targeting zinc uptake in bacteria.

Challenges and Future Directions

Despite the incredible progress in understanding membrane transporters, challenges persist. Many transporters are inherently dynamic, and capturing their conformational changes is a formidable task. Moreover, the scarcity of high-resolution structures for some transporters and their complexes with ligands hampers drug design efforts.

The future of membrane transporter research lies in integrating structural insights with functional and computational studies. Advances in cryo-EM and NMR will enable the study of more dynamic transporters and their interactions with ligands. Additionally, the emergence of hybrid methods that combine multiple structural biology techniques promises to provide a more comprehensive view of transporter mechanisms.

Membrane transporters, with their diverse functions and pivotal roles in health and disease, continue to captivate researchers in the field of structural biology. The quest to decipher their structures and unravel their transport mechanisms has not only deepened our understanding of cellular processes but also holds great promise in the development of novel therapeutics. As technology and methodologies advance, the future of membrane transporter research shines brighter, promising further revelations and innovative solutions to the complex puzzles they present.

18.2 G-Protein Coupled Receptors (GPCRs)

The intricate web of signalling networks within living organisms relies heavily on a class of membrane proteins known as G-Protein Coupled Receptors (GPCRs). These proteins, often referred to as the "molecular switches" of the cell, play a pivotal role in transmitting signals from the extracellular environment to the intracellular realm. While the structural biology of GPCRs has long remained an enigmatic pursuit, recent breakthroughs in the field have illuminated the structure-function relationship of these receptors, opening new avenues for drug discovery and the treatment of various ailments.

Introduction to G-Protein Coupled Receptors (GPCRs)

G-Protein Coupled Receptors are a diverse family of membrane proteins that span the cell membrane. They are responsible for translating a wide range of extracellular signals into intracellular responses. Examples of GPCR-activated signalling pathways include those initiated by neurotransmitters, hormones, and sensory stimuli. A remarkable aspect of GPCRs is their ubiquity; they constitute the largest protein family involved in signal transduction, with over 800 members in the human genome. The widespread significance of GPCRs in human physiology makes them appealing targets for pharmaceutical intervention.

Structural Complexity of GPCRs

The structural complexity of GPCRs has posed a significant challenge for researchers seeking to unravel their secrets. Unlike some membrane proteins with well-defined transmembrane domains, GPCRs have a distinctive architecture. They possess seven transmembrane helices connected by extracellular and intracellular loops. This seven-helix bundle structure is held together by a network of hydrogen bonds, disulfide bridges, and hydrophobic interactions. In the extracellular loops, sites for ligand binding exist, while the intracellular loops engage with G proteins to initiate downstream signalling cascades. This intricate architecture is what gives GPCRs their extraordinary functional diversity.

A Glimpse into GPCR Activation

Understanding GPCR activation is akin to deciphering a molecular Morse code. When a ligand (e.g., a hormone or neurotransmitter) binds to the receptor's extracellular domain, it triggers a conformational change in the receptor. This conformational shift is analogous to flipping a switch, initiating the downstream signalling process. GPCRs are often likened to locks with ligands serving as keys. The binding of the correct ligand induces the necessary conformational change, leading to a specific signalling response. However, the specifics of this conformational change were long shrouded in mystery until the advent of structural biology techniques.

Groundbreaking Structural Insights

The quest to unravel the structures of GPCRs was met with numerous challenges. GPCRs are notoriously hydrophobic, making them difficult to express and purify. Additionally, their flexibility and the instability of many purified GPCR samples have hindered structural studies. Nevertheless, perseverance in the field of structural biology has led to groundbreaking insights.

One of the most iconic milestones in GPCR structural biology was the determination of the crystal structure of rhodopsin, a GPCR involved in visual signal transduction. This achievement marked a turning point, as it provided the first glimpse of a GPCR's three-dimensional structure. Rhodopsin's structure revealed its seven transmembrane helices and the crucial retinal molecule responsible for its light-sensing function. It also shed light on the structural features that are conserved among many GPCRs, such as the "DRY" motif in the second intracellular loop.

However, it wasn't until 2007 that the field witnessed a seismic shift with the resolution of the crystal structure of the β2-adrenergic receptor, a GPCR that responds to adrenaline and noradrenaline. This structure illuminated the complex interplay between the receptor and its bound G protein, giving researchers unprecedented insights into the activation process. The β2-

adrenergic receptor structure confirmed that GPCRs undergo significant conformational changes upon ligand binding and activation. Importantly, it showed how these changes are propagated through the receptor, ultimately leading to G protein activation.

The crystal structures of other GPCRs, such as the adenosine A2A receptor and the adrenergic receptor family, further expanded our knowledge of the GPCR landscape. These structures revealed the diversity of ligand-binding sites and the structural basis for ligand selectivity among GPCRs. Understanding the nuances of ligand binding and activation is paramount for drug design targeting GPCRs.

Significance for Drug Discovery

The structural insights into GPCRs have immense implications for drug discovery. GPCRs are involved in a myriad of physiological processes and are implicated in numerous diseases, making them attractive targets for pharmaceutical research. Understanding the structural details of GPCRs allows for the rational design of drugs that can modulate their activity.

For instance, the β2-adrenergic receptor structure has been pivotal in the development of drugs for conditions like asthma and chronic obstructive pulmonary disease (COPD). By designing ligands that target specific regions of the receptor, researchers can fine-tune drug selectivity and efficacy. This targeted drug design has led to the creation of beta-adrenergic agonists and antagonists that precisely regulate receptor activity.

Moreover, the structural knowledge of GPCRs has spurred the development of biased ligands. These ligands can activate or inhibit specific downstream signalling pathways, offering a more refined approach to drug therapy. For example, a biased ligand for a GPCR involved in pain modulation could provide pain relief without triggering unwanted side effects.

The Role of GPCRs in Disease

The pivotal role of GPCRs in human physiology also extends to their involvement in various diseases. Malfunctions in GPCR signalling can lead to pathological conditions. For instance, mutations in GPCRs involved in vision can cause retinal degenerative diseases, while alterations in adrenergic receptors have been linked to heart diseases and hypertension.

Structural insights into GPCRs have shed light on the molecular underpinnings of these diseases. This knowledge enables the development of precision medicines that target the specific molecular defects associated with GPCR-related conditions. As our understanding of GPCR structures deepens, we can anticipate more effective treatments with fewer side effects.

Future Prospects and Challenges

The journey of GPCR structural biology is far from over. While the past two decades have witnessed remarkable progress, many GPCRs remain unstudied due to challenges in expression, stability, and crystallization. Additionally, there is a growing interest in understanding the dynamics of GPCRs in their native cellular environments, which can be achieved through advanced techniques like cryo-electron microscopy and NMR spectroscopy.

Furthermore, GPCR structural biology is expanding to explore the intricacies of receptor complexes, including those with arrestins and other signalling partners. Understanding these interactions at the atomic level will provide a more comprehensive picture of GPCR signalling.

The structural biology of GPCRs has emerged as a cornerstone in modern drug discovery and our comprehension of diseases linked to GPCR dysfunction. The remarkable progress made in this field not only highlights the importance of perseverance in scientific activities but also exemplifies the potential of structural biology to transform the pharmaceutical landscape. GPCRs, once enigmatic, are now within the realm of understanding, offering new hope for the development of targeted therapeutics and personalized medicine. As the needlepoint of GPCR structural biology continues to unfold, it promises to weave a brighter future for human health and well-being.

18.3 Ion Channels

Ion channels, integral membrane proteins, play a pivotal role in maintaining the electrical potential of cell membranes, a process essential for various physiological functions. They are the microscopic gatekeepers that allow specific ions to traverse the hydrophobic lipid bilayer and regulate a plethora of cellular processes, from muscle contraction to synaptic transmission. This section will provide an in-depth exploration of ion channels, highlighting their structural diversity, functional significance, and notable case studies that have significantly advanced our understanding of these critical proteins.

Introduction to Ion Channels

Ion channels are fundamental components of cell membranes, embedded within the lipid bilayer. They form aqueous pores that selectively facilitate the movement of ions across the membrane.

The diversity of ion channels enables the exquisite control of ion flow, which is crucial for maintaining cellular homeostasis and signalling. These channels are classified based on the types of ions they transport, including sodium (Na+), potassium (K+), calcium (Ca2+), and chloride (Cl-), among others.

One of the key features that makes ion channels fascinating subjects of study is their specificity and selectivity. Each channel typically allows only one or a few types of ions to pass, with remarkable discrimination among closely related ions. This high specificity arises from the precise arrangement of amino acids within the channel's pore, creating an environment that favours the movement of specific ions while hindering others.

Structural Insights into Ion Channels

To gain insights into the structure and function of ion channels, researchers have employed various techniques, including X-ray crystallography, cryo-electron microscopy (cryo-EM), and nuclear magnetic resonance (NMR) spectroscopy. These studies have illuminated the atomic-level details of ion channel architecture.

Potassium Channels (K+): Perhaps one of the most extensively studied ion channels is the potassium channel. The 2003 Nobel Prize in Chemistry was awarded to Rod MacKinnon for his work on elucidating the structure of a bacterial K+ channel, known as KcsA. This breakthrough unveiled the basic architecture of potassium channels, characterized by a central pore formed by a tetrameric arrangement of subunits. These channels display remarkable selectivity for potassium ions, due to the presence of specific binding sites within the pore that coordinate K+ ions while excluding other ions.

Sodium Channels (Na+): Sodium channels, vital for the initiation and propagation of action potentials in neurons and muscle cells, have been another subject of intensive study. The structure of the bacterial sodium channel, NaChBac, has provided critical insights into the mechanism of ion permeation. Research on voltage-gated sodium channels, like Nav1.4, has illuminated the intricate architecture of the voltage-sensing domains and the pore module. These structures have deepened our understanding of the rapid, voltage-dependent sodium ion flux underlying nerve impulses.

Calcium Channels (Ca2+): Calcium channels regulate Ca2+ influx in response to membrane depolarization and play pivotal roles in neuronal signalling, muscle contraction, and exocytosis. The cryo-EM structure of the L-type calcium channel Cav1.1, a giant protein complex, was determined at near-atomic resolution, revealing intricate details of its transmembrane core and cytoplasmic domains. Such structural data are instrumental in understanding the mechanisms that govern calcium channel function and regulation.

Notable Case Studies

Several case studies in the structural biology of ion channels have advanced our understanding of their functionality and the development of therapeutic agents. Here, we'll delve into a few of these influential case studies:

Kv1.2-2.1: A Potassium Channel Gating Mechanism: The structure of Kv1.2-2.1, a voltage-gated potassium channel, illuminated the intricate process of channel gating. It revealed the structural changes associated with voltage-sensor movement and pore opening. This groundbreaking work has provided the basis for understanding the fundamental principles of ion channel gating mechanisms.

TRPV1: The Capsaicin Receptor: The transient receptor potential vanilloid 1 (TRPV1) channel, which is activated by capsaicin and heat, has garnered considerable attention. Cryo-EM studies of TRPV1 have elucidated the structural changes underlying channel activation and the binding of capsaicin. This structural insight has implications for pain perception and offers a potential target for pain management therapies.

NavAb: A Prokaryotic Sodium Channel: The structural analysis of NavAb, a prokaryotic sodium channel, has significantly contributed to our understanding of the sodium channel family. It helped clarify the structural features that allow sodium channels to selectively conduct sodium ions and provided insights into the mechanisms of fast inactivation, a critical aspect of sodium channel function.

RyR1: The Ryanodine Receptor: While not a traditional ion channel, the ryanodine receptor (RyR1) plays a crucial role in calcium release from the sarcoplasmic reticulum in muscle cells. Cryo-EM studies of RyR1 have unveiled its enormous size and complex structure. These findings have deepened our understanding of how this calcium release channel is regulated and its role in muscle contraction.

Therapeutic Implications

Understanding the structure and function of ion channels has profound implications for drug development and therapeutics. Many diseases are associated with ion channel dysregulation, making them attractive targets for drug intervention. For example:

Voltage-Gated Sodium Channels in Channelopathies: Mutations in sodium channels can lead to channelopathies, such as inherited cardiac arrhythmias and epilepsy. Drugs that selectively target sodium channels can be used to modulate their activity, offering treatment options for these conditions.

Calcium Channels in Cardiovascular Disease: Calcium channel blockers, which modulate the activity of calcium channels, are widely used to treat conditions like hypertension and angina. Understanding the structural basis of calcium channel function aids in the development of more effective drugs.

Potassium Channels in Cancer: Some cancer cells exhibit altered potassium channel activity. Targeting these channels with specific inhibitors can be a strategy for cancer therapy. Knowledge of potassium channel structure helps in designing targeted drugs.

Future Directions

The field of ion channel structural biology continues to evolve, with several exciting avenues for future research. As technology advances, researchers can explore ion channels in increasingly complex biological environments. Moreover, the study of ion channel dynamics and the interactions with other membrane proteins and lipids will be crucial for a more comprehensive understanding of their function.

Advanced techniques, such as time-resolved structural studies and single-molecule spectroscopy, will provide new insights into the dynamic aspects of ion channel function. Additionally, integrating structural data with computational modelling will enable the prediction of channel behaviour under various conditions and aid in the development of novel drugs targeting ion channels.

Ion channels are remarkable membrane proteins with diverse structures and functions. The advances in structural biology have shed light on their intricate mechanisms and opened the door to potential therapeutic interventions. As research in this field continues, we can anticipate even greater revelations that will impact our understanding of cellular physiology and the development of treatments for a range of diseases.

18.4 Photosynthetic Membrane Proteins

Photosynthesis, one of the most fundamental processes on Earth, sustains life by converting solar energy into chemical energy. At its heart are photosynthetic membrane proteins, remarkable molecular machines embedded within the thylakoid membranes of plant chloroplasts and bacterial cell membranes. These proteins drive the intricate dance of capturing, converting, and storing solar energy. The structural elucidation of photosynthetic membrane proteins has not only deepened our understanding of their function but has also illuminated the incredible adaptability of life to harness energy from the sun. In this section, we explore the fascinating world of photosynthetic membrane proteins, highlighting key examples and their structural insights.

Photosynthesis: A Masterpiece of Energy Conversion

Photosynthesis, often termed "the green engine of life," is a meticulously choreographed process. It enables plants, algae, and certain bacteria to harvest sunlight and transform it into the chemical energy stored in the form of adenosine triphosphate (ATP) and reduced nicotinamide adenine dinucleotide phosphate

(NADPH). These energy-rich molecules subsequently fuel the synthesis of sugars and other essential compounds, supporting the growth and survival of organisms. The heart of photosynthesis lies in photosynthetic membrane proteins, which perform the astonishing task of converting solar energy into chemical energy.

Photosystem I (PSI): The Primordial Photoconverter

At the forefront of solar energy conversion stands Photosystem I (PSI). This photosynthetic membrane protein complex is the oldest and most conserved component in the photosynthetic machinery, with its origins dating back over a billion years. PSI is central to the light-dependent reactions of photosynthesis, which transpire in the thylakoid membrane of chloroplasts in eukaryotes and the plasma membrane of photosynthetic bacteria. This remarkable protein complex is responsible for the reduction of NADP+ to NADPH and the pumping of protons across the membrane, contributing to the establishment of the proton motive force.

Structure of Photosystem I

Structural studies have unveiled the intricate architecture of PSI. It comprises a multitude of protein subunits, cofactors, and pigments, all working in unison to capture and convert solar energy. The core of PSI is a heterodimeric complex, PsaA and PsaB, each containing over 10 transmembrane helices. Embedded within this core are several cofactors, most notably chlorophylls and carotenoids, which play crucial roles in light absorption and energy transfer. The reaction centre of PSI is home to a pair of chlorophyll a molecules, termed P700, which are responsible for the initial photochemistry. As light is absorbed, these chlorophylls become electronically excited, initiating a cascade of electron transfer reactions.

Electron Transfer in PSI

The electron transfer process in PSI is a remarkable feat of biological engineering. As P700's excited electrons are ejected, they enter an intricate network of iron-sulphur clusters, with the terminal electron acceptor being ferredoxin. These electron transfers are highly efficient, resulting in the reduction of NADP+ to NADPH. PSI also plays a pivotal role in generating a proton gradient across the membrane, contributing to ATP synthesis. Notably, the detailed structure of PSI has illuminated how these electron transfers occur without energy loss, thanks to precise spatial organization and protein-bound cofactors.

Photosystem II (PSII): Splitting Water and Liberating Oxygen

While PSI initiates electron flow, it is Photosystem II (PSII) that introduces a radical departure from conventional wisdom by enabling the oxidation of water. PSII is a large, multisubunit complex located in the thylakoid membrane. It is responsible for the photolysis of water, generating electrons, protons, and oxygen. This oxygenic photosynthesis is the source of molecular oxygen, a vital component for aerobic life on Earth.

The Catalytic Reaction in PSII

The catalytic reaction within PSII is among the most remarkable in nature. It involves the splitting of water into oxygen and protons, releasing electrons in the process. This reaction occurs at a metalloenzyme known as the oxygen-evolving complex (OEC), situated within PSII. The OEC consists of four manganese ions, one calcium ion, and several oxygen atoms. It acts as a water-oxidizing catalyst, providing a platform for water oxidation. High-resolution structural studies have revealed the exact arrangement of these metal ions and the involvement of nearby amino acid residues. The elucidation of the OEC's structure has been an exemplar of how structural biology can illuminate complex catalytic mechanisms in membrane proteins.

PSII's High-Resolution Structure

Studying PSII's structure is a milestone in understanding the workings of photosynthetic membrane proteins. Recent advances in cryo-electron microscopy have made it possible to determine PSII's structure at near-atomic resolution. These studies have provided valuable insights into the arrangement of protein subunits, cofactors, and the catalytic centre. PSII's core complex comprises over 20 protein subunits, and this structural information has helped clarify the role of each subunit in electron transfer, proton pumping, and stabilizing the complex.

Cytochrome b6f Complex: The Energy Transformer

While PSI and PSII play pivotal roles in electron transfer, the Cytochrome b6f complex bridges the two photosystems, shuttling electrons and generating a proton gradient. This remarkable protein complex, residing in the thylakoid membrane, couples electron transfer to the pumping of protons across the membrane.

Electron Transport in Cytochrome b6f Complex

The Cytochrome b6f complex serves as an electron conduit, transferring electrons from PSII to PSI. As electrons move through this complex, protons are pumped across the membrane, contributing to the proton motive force. A fascinating aspect of this complex is its ability to harness the energy released during electron transfer to perform mechanical work, such as rotating subunits. The energy transformation within the Cytochrome b6f complex is a testament to the adaptability of membrane proteins in capturing and utilizing energy.

High-Resolution Insights

In recent years, the high-resolution structures of the Cytochrome b6f complex have been determined using techniques such as X-ray crystallography and cryo-electron microscopy. These structures have unveiled the precise arrangement of protein subunits and the presence of redox-active cofactors. Notably, they have also illuminated the mechanism of proton pumping, shedding light on how the complex couples electron transfer to the transport of protons across the membrane.

Light-Harvesting Complexes: Antennae of Photosynthesis

Light-harvesting complexes (LHCs) are intricate assemblies of proteins and pigments that play a crucial role in capturing and channelling light energy to the reaction centres of PSI and PSII. These complexes act as antennas, expanding the range of light absorption and protecting the photosystems from photodamage.

The Structure of LHCs

LHCs are highly dynamic and adaptable, with variable numbers of chlorophylls, carotenoids, and xanthophylls in their binding sites. The structures of LHCs have been challenging to determine due to their flexibility and the presence of multiple pigment-binding sites. However, recent advances in structural biology have allowed for the elucidation of LHC structures. These studies have provided insights into how LHCs capture and transfer light energy, offering a glimpse into the rapid and efficient energy transfer mechanisms at play.

Role in Photoprotection

Beyond light harvesting, LHCs also play a vital role in photoprotection. In high-light conditions, they can dissipate excess absorbed energy as heat, preventing photodamage to the photosystems. The structural elucidation of LHCs has shed light on the mechanisms behind this photoprotective role, highlighting the importance of these complexes in the adaptability of photosynthetic organisms to varying light conditions.

The Continuing Saga of Photosynthetic Membrane Proteins

The structural biology of photosynthetic membrane proteins has brought us closer to unravelling the mysteries of one of life's most awe-inspiring processes. These proteins, residing in the membranes of chloroplasts and photosynthetic bacteria, capture and convert solar energy with remarkable efficiency. The structures of PSI, PSII, the Cytochrome b6f complex, and LHCs have provided profound insights into the mechanisms underlying photosynthesis.

In the future, continued research in this field will likely uncover further details, enabling us to comprehend the intricate details of energy conversion in photosynthetic organisms. Moreover, as we seek sustainable energy solutions and a deeper understanding of life's adaptability, the lessons learned from photosynthetic membrane proteins may illuminate novel approaches to harnessing solar energy and provide inspiration for a more sustainable future.

Chapter 19: Future Directions in Membrane Protein Research

19.1 Current Challenges and Unanswered Questions

In the exhilarating voyage through the structural biology of membrane proteins, we have unearthed a treasure trove of insights into the intricacies of life at the molecular level. These molecules, studded in the cellular fabric, are pivotal players in an enigmatic dance of biological processes. However, as we navigate this expansive sea of knowledge, we encounter not just well-charted waters but also uncharted territories, where tempests of uncertainty brew. In this section, we don our metaphorical captain's hats and set sail to explore the currents, the challenges, and the mysteries that continue to captivate the hearts and minds of membrane protein researchers.

The Challenging World of Membrane Protein Expression

Before the era of high-resolution structural techniques, membrane protein researchers faced the daunting challenge of obtaining sufficient quantities of these proteins for study. While considerable progress has been made, this challenge remains relevant. Producing enough membrane proteins for structural studies is akin to a mariner in need of a constant supply of fresh water. One cannot understate the significance of a bountiful harvest.

The variability in success across different membrane proteins is perplexing. Some proteins express in abundance, while others prove as elusive as the fabled Fountain of Youth. Overexpression, a common technique for obtaining membrane proteins, may trigger a cascade of issues, such as misfolding, aggregation, or poor solubility. The expression system chosen can also impact yield, with some proteins thriving in bacterial hosts, while others demand the intricate machinery of eukaryotic cells.

Let us take the case of the G-protein-coupled receptors (GPCRs) as an example. These ubiquitous membrane proteins play a pivotal role in signal transduction, making them attractive drug targets. Despite their significance, GPCR overexpression is often beset with challenges. An elusive treasure, GPCRs are notorious for their low expression levels, complex folding, and susceptibility to misfolding. Researchers have turned to various expression systems, from yeast to insect cells, to overcome these hurdles. Yet, it remains a quest to consistently obtain high-quality, well-folded GPCR samples.

Membrane Protein Dynamics: A Whisper in the Dark

The structure of a membrane protein, akin to a photograph, captures a single moment in its existence. But like a photograph, it fails to convey the bustling life within. Understanding a membrane protein's dynamics is akin to discerning the movement of ships under a moonless night sky. The intrinsic flexibility of these proteins, their conformational changes, and their interactions with lipids, ligands, and other proteins remain an enigma.

While X-ray crystallography and cryo-electron microscopy (Cryo-EM) have bestowed detailed snapshots, these static images do not narrate the full story. NMR spectroscopy has made strides in unveiling these dynamics, but it faces challenges with larger proteins and complex samples.

An exemplary challenge in membrane protein dynamics lies in the world of transporters. These proteins facilitate the movement of ions, metabolites, and signalling molecules across the membrane. Understanding their conformational changes during transport is pivotal. The recent structural elucidation of transporters like SLC26 anion exchangers and SLC22 drug transporters marks significant progress. Still, the question of precisely how these transporters transition between states remains a puzzling one.

Elusive Transient Interactions

Proteins are seldom loners in the cellular milieu. Membrane proteins engage in intricate interactions with lipids, co-factors, ions, and other proteins. These interactions underpin a wide array of biological processes, from signal transduction to membrane transport. Yet, they are as challenging to capture as fireflies in the night.

Take the case of membrane proteins that partner with lipids. Understanding the specific lipid-protein interactions that dictate membrane protein function and stability is akin to identifying constellations in a cityscape drowned in light pollution. Various techniques like mass spectrometry and nuclear magnetic resonance (NMR) provide glimpses, but a comprehensive picture remains elusive.

Similarly, studying protein-protein interactions, a key component of cell signalling and regulation, is a quagmire of complexity. GPCRs, for instance, are known to interact with an array of intracellular partners, such as G proteins and arrestins.

Mapping these interactions requires innovative approaches, as the interactions are often transient and dynamic. It's akin to deciphering a bustling market square, where individuals continually change partners in a lively dance.

The Role of Membrane Proteins in Disease: The Unsolved Riddle

Membrane proteins are central actors in various diseases, making them prime targets for therapeutic intervention. For example, ion channels are associated with conditions such as epilepsy and cardiac arrhythmias, while mutations in transporters lead to inherited metabolic disorders. GPCRs, on the other hand, are implicated in diverse pathologies, including cancer and neurological diseases.

Despite their clinical significance, the precise mechanisms by which these proteins contribute to disease often remain uncertain. Deciphering the link between a specific genetic mutation and the resulting pathological phenotype is akin to navigating a maze. The conformational dynamics and ligand interactions of disease-relevant membrane proteins are intricately tied to disease mechanisms. Yet, these connections remain elusive in many cases, and pharmaceutical interventions often target symptoms rather than underlying causes.

The Next Horizon: Emerging Technologies

As we sail through the challenges and unanswered questions in membrane protein research, a glimmer of hope appears on the horizon. Emerging technologies promise to shed new light on these mysteries.

Nanodisc technology, for instance, offers a novel approach to membrane protein stabilization, enabling a broader range of proteins to be studied. Lipidic cubic phase techniques continue to evolve, promising improved crystallization of membrane proteins. Advanced labelling techniques, such as site-specific isotopic labelling, hold the potential to enhance NMR studies of larger membrane proteins.

Furthermore, the burgeoning field of artificial intelligence and machine learning is poised to transform our ability to analyse complex data and predict protein structures. These technologies hold the promise of accelerating our journey through the sea of membrane protein mysteries.

Therefore, the structural biology of membrane proteins has made significant strides, but the voyage is far from over. The challenges and unanswered questions in this field are like the vast expanse of an ocean, and we are but intrepid explorers setting forth on a never-ending quest for knowledge. With each challenge we overcome and each mystery we unravel, the allure of these proteins grows stronger, propelling us further into the captivating world of the cell's membrane-bound sentinels.

19.2 Integration with Systems Biology

The world of membrane proteins is a veritable treasure trove of biological insights. The structural biology of these proteins, as explored in the preceding chapters, has provided an indispensable foundation for comprehending their intricate roles in cellular processes. Yet, a broader vista beckons us—a horizon where individual proteins are not viewed in isolation but as integral components of the larger biological picture. This section delves into the integration of membrane protein research with the broader discipline of systems biology, a dynamic marriage that promises to revolutionize our understanding of complex biological systems.

Understanding the Interplay of Proteins

While the structural elucidation of individual membrane proteins has unveiled their secrets on a molecular scale, the need to understand how these proteins collaborate within the context of a living cell has long been recognized. Systems biology, at its core, seeks to decipher the intricate web of interactions between biomolecules—proteins, nucleic acids, and metabolites—across a wide spectrum of biological processes. It's the scientific equivalent of putting together the pieces of a sprawling jigsaw puzzle, where the membrane proteins are but one piece of the grand mosaic.

One of the quintessential questions that systems biology endeavors to answer is how membrane proteins participate in complex cellular networks. Consider, for example, a G-protein coupled receptor (GPCR) that orchestrates cellular responses to extracellular signals. While structural biology might yield a detailed portrait of a GPCR's atomic structure, it remains silent about the cascade of events triggered upon ligand binding. Systems biology, on the other hand, steps into the fray, aiming to decipher the signalling pathways, secondary messengers, and eventual cellular responses that ensue. This collaborative approach is essential for a comprehensive understanding of how membrane proteins function in their natural habitat.

Data Integration: A Complex Needlepoint of Information

Central to the integration of structural biology and systems biology is the harmonization of diverse datasets. Structural biologists are adept at generating high-resolution models of individual proteins, but the complexity of biological systems demands a more holistic approach. Integrating structural data with functional, proteomic, and transcriptomic information is akin to weaving a complex needlepoint, where each thread represents a different facet of cellular biology.

One of the challenges in this endeavour is the diversity of data types. For instance, one dataset might comprise the three-dimensional structure of a membrane protein, while another could be a time-series of gene expression data under varying conditions. To illustrate, researchers studying a membrane protein involved in cell signalling would be keen to understand how its expression varies across different cell types and in response to various ligands. This, in turn, necessitates the integration of structural information with data on gene expression, protein-protein interactions, and functional assays, presenting an intricate challenge that systems biologists readily embrace.

Computational Approaches and Modelling

The integration of structural and systems biology is underpinned by a suite of computational approaches and modelling techniques. These are indispensable tools that allow researchers to bridge the gap between the atomic-scale insights afforded by structural biology and the broader systemic context provided by systems biology.

For example, the use of molecular dynamics simulations can extend our understanding of how membrane proteins interact with lipids and other cellular components. These simulations, when applied to a lipid bilayer, can elucidate the dynamics of membrane protein-lipid interactions, providing insights into the protein's behaviour in a near-physiological environment. Such simulations enable researchers to infer how specific structural features, like the positioning of transmembrane helices, influence a protein's function.

Another critical aspect of computational modelling lies in the prediction of protein-protein interactions. Systems biology relies heavily on the identification of such interactions to build comprehensive network maps. Structural data, in this case, plays a crucial role in understanding the spatial arrangements of interacting partners. As an example, structural information about the extracellular domains of membrane proteins can be pivotal in predicting their interactions with extracellular ligands, neighbouring proteins, or downstream signalling components.

Quantitative Data and Network Analysis

In the realm of systems biology, the emphasis is on quantitative data, and the integration of structural information elevates the precision and accuracy of models. For instance, the stoichiometry of protein complexes and the binding affinities between interacting partners can be informed by structural data. In the context of membrane proteins, this becomes especially pertinent when considering the formation of multimeric complexes. The structural details of a receptor-ligand interaction, coupled with kinetic data, can lead to more accurate predictions of signal transduction dynamics.

Furthermore, network analysis techniques are instrumental in dissecting the interactions between membrane proteins and other cellular components. By integrating structural data with quantitative measurements of protein abundance and activity, researchers can construct network models that provide a bird's-eye view of the cellular machinery. These models reveal hubs of interaction, signalling pathways, and potential points of intervention, which are invaluable for understanding disease mechanisms or designing therapeutic strategies.

Case Study: GPCR Signalling Network

To illustrate the synergy of structural and systems biology, consider the example of GPCR signalling, a highly intricate process involved in cell communication and response to extracellular signals. Structural biology has provided an abundance of GPCR crystal structures, shedding light on the conformational changes these receptors undergo upon ligand binding. However, it's systems biology that illuminates the signalling cascade initiated by GPCRs.

The structural details of GPCR activation, such as the ligand-induced conformational changes, are crucial pieces of the puzzle. Integrating this structural information with data on downstream protein-protein interactions, enzymatic activities, and gene expression profiles reveals the full scope of GPCR signalling. This comprehensive approach has unravelled intricate signalling networks, including GPCR-mediated activation of intracellular kinases and the subsequent modulation of gene expression.

Moreover, such integration has therapeutic implications. By pinpointing critical nodes in the GPCR signalling network, researchers can identify potential drug targets and develop more effective pharmaceutical interventions. In this way, the union of structural and systems biology leads to a deeper understanding of cellular processes and provides a roadmap for therapeutic development.

Challenges and Future Directions

The integration of structural biology with systems biology is a dynamic and evolving field, but it is not without its challenges. First and foremost is the need for standardized data sharing and collaboration. Researchers must work together to create interoperable databases and develop common data formats that facilitate the exchange of information between structural and systems biology communities.

Another challenge lies in the development of predictive models that incorporate structural data seamlessly. While structural information can inform these models, it is essential to ensure that predictions align with experimental observations. This requires iterative refinement and validation of computational models.

In the future, as structural biology techniques continue to advance, the integration with systems biology will become even more intricate. Cryo-electron microscopy, for instance, is revolutionizing our ability to study membrane proteins in their native lipid environments, opening new avenues for data integration and network analysis.

Henceforth, the integration of structural biology with systems biology represents a paradigm shift in our approach to understanding membrane proteins. It allows us to move beyond the confines of isolated structures and explore the vast landscape of biological networks. As we continue to weave together structural, functional, and network data, we will unravel the needlepoint of life's complexity, revealing the inner workings of the cellular machinery, and providing insights that are essential for both basic science and the development of novel therapeutics.

The future of membrane protein research lies at the intersection of these two complementary disciplines, where exciting discoveries and breakthroughs await.

19.3 Potential Technological Breakthroughs

In the ever-advancing field of membrane protein research, the landscape of possibilities is continually shifting. As we navigate the intricacies of this dynamic terrain, it is essential to examine the technological breakthroughs that are reshaping the future of our understanding of these crucial biomolecules. In this section, we will explore the most promising advancements, from cutting-edge techniques to innovative tools that promise to revolutionize our approach to studying membrane proteins.

Cryogenic Electron Microscopy (Cryo-EM): Redefining Structural Resolution

Cryo-EM has experienced a resurgence over the last decade, emerging as a revolutionary tool in the structural biology toolbox. Its applications have extended beyond soluble proteins to membrane proteins, where traditional techniques often faltered. Cryo-EM has the unique advantage of studying biological specimens in their near-native state, eliminating the need for crystallization and preserving the structural integrity of membrane proteins.

One of the most remarkable breakthroughs in Cryo-EM is the development of direct electron detectors. These detectors have improved the signal-to-noise ratio, enabling researchers to capture high-resolution images of membrane protein complexes. For instance, the structure of the human glutamate receptor, a multi-subunit membrane protein complex involved in synaptic transmission, was elucidated at atomic resolution using Cryo-EM. This achievement underscores the potential of Cryo-EM to uncover the intricate details of membrane protein structures, furthering our understanding of their function and pharmacological potential.

Nanodisc Technology: Stabilizing Membrane Proteins in a Native-Like Environment

Nanodiscs are nanoscale lipid bilayer patches encircled by amphipathic membrane scaffold proteins (MSPs). This technology is a game-changer for membrane protein research, providing a native-like environment for these proteins while offering stability and solubility. Nanodiscs have the unique ability to encapsulate a single membrane protein or even an entire complex, allowing for detailed structural and functional investigations.

An exciting development in this field is the utilization of designer lipids and MSPs. These enable researchers to precisely tailor the lipid composition and structural properties of nanodiscs, providing a customized environment for specific membrane proteins. For example, the introduction of charge-altering mutations in MSPs has facilitated the controlled assembly of nanodiscs, expanding the scope of this technology.

Single-Particle Analysis: A Leap Forward in Dynamics and Flexibility

Historically, membrane protein research has often struggled to capture the dynamics and flexibility of these biomolecules, which are crucial to their function. However, recent advances in single-particle analysis have the potential to bridge this gap. This technique involves the analysis of individual protein complexes and their variations, offering insights into their conformational changes and functional states.

In the case of G-protein coupled receptors (GPCRs), single-particle analysis has been instrumental in revealing distinct conformational states. By examining individual receptor molecules, researchers have gained a deeper understanding of GPCR activation and signalling. This knowledge holds immense promise for the development of more targeted pharmaceuticals aimed at these critical cell membrane receptors.

Advanced Labelling Strategies: Shedding Light on Protein Dynamics

One of the enduring challenges in membrane protein research has been monitoring the dynamics of these proteins in their native environment. To overcome this hurdle, advanced labelling strategies have emerged, providing a means to track specific regions or residues within the protein.

One such breakthrough is the development of site-specific labelling techniques. For example, researchers can use amber codon suppression to introduce non canonical amino acids into the protein, allowing for selective labelling. This strategy has enabled the study of dynamic regions in membrane proteins, shedding light on their conformational changes during various biological processes.

Hydrogen-Deuterium Exchange Mass Spectrometry (HDX-MS): Probing Structural Dynamics

HDX-MS is a powerful technique that has gained traction in membrane protein research by probing structural dynamics. It involves the measurement of the exchange of hydrogen atoms with deuterium in the protein backbone, providing information about its solvent accessibility and conformational changes.

In the context of membrane proteins, HDX-MS has been applied to study the dynamics of integral membrane proteins. By comparing the deuterium exchange rates of different regions, researchers can identify flexible regions and monitor changes induced by ligand binding or membrane interactions. This information is invaluable for understanding the function and regulation of membrane proteins.

Interdisciplinary Approaches: A Collaborative Outlook

As we delve further into the realm of membrane protein research, it is clear that the most significant breakthroughs often emerge at the intersection of multiple disciplines. Collaborative efforts between structural biologists, computational scientists, chemists, and biophysicists are becoming increasingly common. These multidisciplinary teams bring together diverse expertise, driving innovation and addressing complex questions in membrane protein research.

For instance, the combination of computational modelling and experimental data has become a powerful approach. Molecular dynamics simulations can complement experimental findings by providing atomistic insights into the dynamics of membrane proteins. By fusing these approaches, researchers gain a more comprehensive understanding of membrane protein behaviour.

In the dynamic and ever-evolving landscape of membrane protein research, promising technological advances continue to redefine the boundaries of what we can achieve. Cryo-EM, nanodisc technology, single-particle analysis, advanced labelling strategies, HDX-MS, and interdisciplinary collaborations are propelling us toward a deeper understanding of membrane proteins. These breakthroughs not only unravel the intricacies of these vital biomolecules but also hold the key to future developments in drug design, therapeutic strategies, and our comprehension of life at the cellular level. The possibilities are as vast as the horizon, and our pursuit of knowledge in this field remains as exciting as ever.

19.4 Interdisciplinary Approaches

In the active and ever-changing arena of membrane protein research, interdisciplinary approaches have become indispensable. These multifaceted strategies harness the collective power of diverse scientific disciplines, forging innovative pathways to explore the complex world of membrane proteins. From combining structural biology with biophysics to integrating bioinformatics with chemistry, these collaborative endeavors illuminate novel perspectives on membrane protein structure and function. This chapter will explore the captivating synergy of interdisciplinary approaches, exemplifying how it shapes the future of this field.

Synergy of Structural Biology and Biophysics

When discussing membrane proteins, the convergence of structural biology and biophysics is paramount. Structural biology offers us glimpses into the three-dimensional architecture of these proteins, while biophysics probes their dynamic behaviour at the atomic and molecular levels. An excellent illustration of this synergy is the study of ion channels, such as the potassium channel (KvAP). In recent years, the combination of X-ray crystallography and electrophysiology has shed light on the intricate mechanisms underlying ion permeation and selectivity. This union has not only revealed the precise arrangement of amino acids but also provided a nuanced understanding of the electrostatic forces governing ion movement through the channel's pore.

Integrating Mass Spectrometry and Proteomics

Mass spectrometry and proteomics have become pivotal in dissecting the intricacies of membrane protein complexes. For example, the mitochondrial respiratory chain, housing a plethora of membrane proteins, is a formidable biological entity. By employing mass spectrometry, scientists can unravel the dynamic interplay within this complex, identifying its constituents and quantifying their abundance. This holistic approach has not only broadened our knowledge of membrane protein interactions but has also contributed to our understanding of how dysregulation within these complexes can lead to various diseases, including mitochondrial disorders.

The Marriage of Bioinformatics and Chemistry

In the realm of membrane proteins, the union of bioinformatics and chemistry is transformative. The task of predicting membrane protein structures has been considerably advanced by computational methods rooted in bioinformatics. Integrating these predictions with chemical strategies, such as covalent labelling and mass spectrometry, enables experimental verification and validation. The rhodopsin family of G protein-coupled receptors (GPCRs) is a prime example. Utilizing bioinformatics predictions in tandem with chemical probing techniques, researchers have deciphered the structural dynamics of GPCRs and how ligand binding induces conformational changes. This convergence of disciplines not only guides drug design but also provides a deeper understanding of signalling processes initiated by these receptors.

Interdisciplinary Insights into Transporters

Membrane transporters play pivotal roles in cellular homeostasis and are essential for nutrient uptake and waste removal. The integration of microbiology and bioinformatics has been instrumental in characterizing transporter proteins. For instance, the bacterial ATP-binding cassette (ABC) transporters are a fascinating subject of study. Combining microbiological growth assays with computational tools has allowed researchers to unravel the substrate specificity and binding mechanisms of ABC transporters, shedding light on how these transporters contribute to antibiotic resistance and virulence in pathogenic bacteria. This collaborative approach underscores the potential for drug development by targeting these transporters.

Confluence of Structural Biology and Cryo-Electron Microscopy (Cryo-EM)

The synergy between X-ray crystallography and cryo-electron microscopy (Cryo-EM) has revolutionized our understanding of membrane protein structures. Take, for example, the case of bacteriorhodopsin, an integral membrane protein found in halophilic archaea. By combining X-ray crystallography with Cryo-EM, researchers achieved high-resolution structures and dynamic snapshots of bacteriorhodopsin, unravelling its photoactive cycle. This interdisciplinary approach allowed scientists to visualize the entire process of retinal isomerization and proton pumping, deepening our comprehension of how membrane proteins perform essential biological functions.

Unveiling Drug Binding Sites through NMR and Molecular Dynamics

Interdisciplinary approaches also have profound implications in drug discovery. The integration of nuclear magnetic resonance (NMR) spectroscopy and molecular dynamics simulations has been instrumental in characterizing drug binding sites in membrane proteins. A prominent example is the study of the HIV-1 protease, a membrane-associated enzyme crucial for viral replication. Through a combination of NMR techniques and molecular dynamics simulations, researchers have elucidated the binding modes of protease inhibitors, providing critical insights for the development of anti-HIV drugs. This integrated approach exemplifies how interdisciplinary strategies can expedite drug discovery processes.

Systems Biology and Membrane Proteins

The rise of systems biology has brought about a holistic perspective on membrane protein research. Systems biology integrates data from genomics, proteomics, and metabolomics, allowing scientists to understand the intricate networks that membrane proteins are embedded in. One remarkable application of this approach is the study of membrane protein interactions in signal transduction pathways. By analysing the entire network of interacting proteins, researchers gain a comprehensive view of how membrane proteins contribute to cellular responses, such as cancer cell proliferation or neuronal signalling.

Ecology Meets Membrane Proteins: Environmental Impact

Interdisciplinary approaches aren't limited to the laboratory setting. Environmental studies have also ventured into the world of membrane proteins. For example, the microbial communities living in extreme environments, like deep-sea hydrothermal vents, are subjected to extreme pressure, temperature, and chemical conditions. Understanding how membrane proteins in these extremophiles function requires an interdisciplinary approach that combines microbiology, geochemistry, and structural biology. Such research not only extends our understanding of the ecological impact of membrane proteins but also provides insights into biotechnological applications in extreme environments.

Ethics and Interdisciplinary Collaborations

While the pursuit of interdisciplinary collaborations is filled with promise, it also necessitates a commitment to ethical considerations. Researchers engaging in interdisciplinary endeavors must navigate issues related to data sharing, intellectual property, and authorship. Additionally, the involvement of human or animal subjects in research mandates adherence to ethical guidelines. As membrane protein research continues to evolve through interdisciplinary approaches, the responsible and ethical conduct of research remains a cornerstone of scientific progress.

In the mission to decrypt the particulars of membrane proteins, the convergence of diverse disciplines has emerged as an indispensable asset. Interdisciplinary approaches provide unique vantage points, allowing researchers to explore new dimensions of these enigmatic biomolecules. As we navigate this ever-evolving terrain, the synergy of structural biology and biophysics, mass spectrometry and proteomics, bioinformatics and chemistry, microbiology and bioinformatics, and other interdisciplinary unions forge a promising path forward. Through these dynamic collaborations, we inch closer to unlocking the secrets of membrane proteins and harnessing their potential for the betterment of science and society. The fusion of knowledge and expertise across disciplines stands as a testament to the immense potential of interdisciplinary approaches in the structural biology of membrane proteins.

Chapter 20: Conclusion and Outlook

20.1 Summary of Key Findings

In this concluding section of our exploration into the structural biology of membrane proteins, we embark on a journey through the intriguing discoveries, breakthroughs, and the transformative impact of this field. As we traverse this path, the knowledge we've gained is not only a testament to the resilience of scientific pursuit but also a beacon lighting the way for future research in this domain. This chapter seeks to unravel, not just the structural intricacies of membrane proteins, but also the incredible implications of our discoveries.

The Proteome Unveiled

The first chapter of this book introduced us to the world of membrane proteins - these enigmatic molecules that are integral to life itself. As we culminate this journey, we cannot help but reflect on the sheer enormity of this realm. It is now understood that approximately one-third of the proteome comprises membrane proteins, underlying their profound significance in biology. Moreover, this fraction plays pivotal roles in various biological processes, from cellular signalling to transport, and energy generation. Our deep-dive into their structures, from X-ray crystallography to cryo-electron microscopy, has unveiled their hidden world, illuminating the mechanisms that drive life as we know it.

The Dichotomy of Challenges and Triumphs

As we explored in Chapter 1, the structural biology of membrane proteins comes with its own set of challenges. Their hydrophobic nature and delicate stability can make them formidable subjects of study. Yet, this challenge has fuelled innovation, and through the annals of history, we've witnessed an interplay of difficulties and triumphs. Notably, overcoming these obstacles led to a parade of innovative methods such as lipidic cubic phase crystallization and advances in detergents for solubilization.

A case in point is the G-protein coupled receptor (GPCR) family, which presented immense hurdles due to their intricate conformational flexibility. However, the dedication of researchers worldwide bore fruit with groundbreaking discoveries like the crystal structure of rhodopsin, which was a watershed moment in structural biology. It highlighted the sheer tenacity and determination of the scientific community in their relentless pursuit of knowledge.

The Revolution of Cryo-Electron Microscopy

The advent of cryo-electron microscopy (Cryo-EM) marked a significant turning point in the field, as we delved into in Chapter 7. The three-dimensional structures of membrane proteins were once considered elusive due to their resistance to crystallization. However, the pioneering work of scientists like Joachim Frank, Jacques Dubochet, and Richard Henderson led to the development of Cryo-EM, offering a more flexible and approachable means of structural determination. This shift was nothing short of revolutionary. Today, researchers can investigate membrane proteins without the constraints of crystallization, allowing a more in-depth understanding of their conformational dynamics and interactions with other biomolecules.

Glimpses into Dynamic Landscapes

Chapter 8 elucidated the dynamic nature of membrane proteins and the tools we employ to uncover their secrets. NMR spectroscopy emerged as a pivotal technique, capable of providing snapshots of their conformational changes and interactions. Through dynamic nuclear polarization, solid-state NMR, and solution NMR, we gained invaluable insights into the roles of membrane proteins in cellular processes. The study of protein-lipid interactions, as discussed in Chapter 9, illuminated the intricate interplay between membrane proteins and the lipid environment, underpinning their structural and functional integrity. Such interactions are not merely structural curiosities but are central to cellular processes such as signal transduction and ion transport.

From Structure to Function

The correlation between structure and function has been a recurring theme throughout this journey. In Chapter 8, we explored how understanding the dynamics of membrane proteins is crucial for comprehending their roles in biological systems. One of the most captivating aspects of membrane proteins is their ability to perform diverse functions. Ion channels, transporters, receptors, and enzymes play pivotal roles in signal transduction, transport of molecules, and energy conversion. For example, the potassium ion channel, discovered through X-ray crystallography, highlighted the precision of membrane protein structures in facilitating ion flux across cellular membranes, enabling vital processes such as muscle contraction and nerve transmission.

The Gateway to Drug Discovery

Chapter 10 shed light on the potential applications of membrane protein research in drug discovery. Membrane proteins are not merely passive participants in cellular activities; they are active targets for pharmacological intervention. GPCRs, in particular, exemplify the success of structure-based drug design. The determination of their structures, such as the beta-adrenergic receptor, paved the way for the development of drugs targeting diverse conditions, from hypertension to asthma. Furthermore, engineering approaches outlined in Chapter 12 provide a roadmap for designing new therapeutic agents and improving drug specificity.

An Ethical Compass

Scientific progress is not devoid of ethical considerations, as elucidated in Chapter 16. We must tread carefully, mindful of the impact our research may have on society and the environment. Ethical considerations extend to the use of animals in research, the sharing of data, and the responsible conduct of research. These are not secondary concerns but are integral to the scientific process. Informed, responsible, and ethical practices ensure that our quest for knowledge does not come at the expense of ethical values.

Future Frontiers and Uncharted Horizons

As we conclude this journey, it is imperative to acknowledge that the structural biology of membrane proteins is an ever-evolving field. The twenty chapters of this book are not the culmination but rather a snapshot of an ongoing exploration. Emerging technologies, as discussed in Chapter 15, such as nanodisc technology and advanced labelling techniques, offer glimpses of exciting prospects. The integration of structural biology with systems biology, as contemplated in Chapter 19, promises a more holistic understanding of cellular processes. The frontier of synthetic biology, wherein we engineer and design membrane proteins, opens up unprecedented possibilities. The future is a canvas upon which we will paint new discoveries, patterns, and intricacies.

The structural biology of membrane proteins is an extraordinary effort that has illuminated the needlepoint of life. We have journeyed from the introduction of these enigmatic molecules to the forefront of scientific innovation. This chapter serves not as a final chapter but as a prologue to the continued quest for knowledge. Our understanding of the structures and functions of membrane proteins is not only a testament to human ingenuity but also a tribute to the perseverance of researchers worldwide. As we pass the torch to future generations, we do so with the conviction that the structural biology of membrane proteins will remain a vibrant and dynamic field, continually shaping our understanding of the molecular underpinnings of life.

20.2 *The Ongoing Impact of Membrane Protein Research*

In the annals of scientific inquiry, few disciplines have witnessed the profound impact and transformative potential of membrane protein research. This frontier of biophysics and structural biology, whose significance cannot be overstated, continues to be an unfolding narrative of exploration, discovery, and application. It is here, in this remarkable journey, that we see the relentless march of progress, changing the course of scientific and medical landscapes.

From Bench to Bedside: Membrane Proteins in Drug Development

One of the most striking impacts of membrane protein research is on drug development. Membrane proteins, often draped in obscurity due to their complex structure and inaccessibility, are, paradoxically, key players in the pharmaceutical industry. As of June 2023, over 50% of all approved drugs target membrane proteins, underscoring their paramount importance in medicine.

Consider G-Protein Coupled Receptors (GPCRs), an illustrious class of membrane proteins. GPCRs, vital in signal transduction, regulate processes as diverse as neurotransmission, immune responses, and hormonal regulation. The 2012 Nobel Prize in Chemistry was awarded to Robert Lefkowitz and Brian Kobilka for their groundbreaking work on GPCRs, exemplifying the transformative power of membrane protein research. The identification and characterization of these receptors have opened doors to novel drugs for conditions ranging from heart disease to depression. Thus, membrane proteins have become the gold mines of pharmacology, revolutionizing how we combat diseases.

Moreover, the advent of structural biology techniques like X-ray crystallography and cryo-electron microscopy has provided the ability to visualize the intricate details of membrane protein structures. This capability has accelerated drug development. For instance, the structural determination of the β2-adrenergic receptor, a GPCR, revealed critical binding sites for drugs treating asthma and chronic obstructive pulmonary disease (COPD). These breakthroughs have not only transformed the pharmaceutical landscape but also underscore the tremendous potential yet to be harnessed as we continue to uncover the structures of more membrane proteins.

Navigating the Uncharted Waters: Membrane Proteins in Health and Disease

The study of membrane proteins extends well beyond drug development, reaching deep into our understanding of health and disease. Membrane proteins are involved in a staggering array of biological processes. Their role in cellular transport, signalling, and homeostasis is indisputable. Researchers have identified mutations in genes encoding membrane proteins that underlie numerous genetic disorders, such as cystic fibrosis and muscular dystrophy.

Cystic fibrosis, for instance, is caused by mutations in the CFTR gene, which encodes a chloride ion channel. Decades of research into CFTR's structure and function have unveiled insights critical for developing therapies. The recently approved drug Ivacaftor is a shining example. It modulates CFTR function, alleviating the symptoms of cystic fibrosis in a significant proportion of patients. Such achievements emphasize the immediate impact of membrane protein research on the lives of those afflicted by devastating diseases.

The scope of membrane proteins in health and disease continues to broaden. In cancer research, membrane proteins are frequently implicated as drivers of tumour development. Receptor tyrosine kinases, which span the cell membrane, are notorious for their roles in uncontrolled cell growth and are prime targets for cancer therapies. The success of drugs like Imatinib in treating chronic myeloid leukaemia by inhibiting the BCR-ABL fusion protein highlights the potential for membrane proteins as therapeutic targets in oncology.

In neurological disorders, the importance of membrane proteins cannot be overstated. The neurodegenerative disease Alzheimer's, for instance, involves the aberrant processing of amyloid precursor protein (APP), a transmembrane protein. The mechanisms underlying the cleavage of APP are an area of intense study in membrane protein research, with the hope of eventually devising therapies to slow or halt the progression of this devastating disease.

The Structural Basis of Transport: Membrane Proteins in Physiology

Membrane proteins are the gatekeepers of life, controlling the flow of molecules across cellular membranes. Transporters, channels, and pumps are among the diverse classes of membrane proteins that ensure the right molecules reach their intended destinations.

Consider the voltage-gated ion channels that orchestrate the transmission of nerve impulses. The pioneering work of Roderick MacKinnon, a Nobel laureate, laid bare the intricate structure of these channels. His research highlighted the role of these channels in diseases like epilepsy and cardiac arrhythmias, leading to a deeper understanding of their mechanisms and the development of more precise therapies.

Membrane proteins are not confined to the eukaryotic realm. Bacterial transporters, such as the lactose permease, have played a pivotal role in the understanding of the transport of molecules across cellular membranes. This knowledge is not only crucial in comprehending bacterial physiology but also in the broader context of antibiotic resistance. The study of membrane proteins in bacteria has illuminated the mechanisms by which bacterial cells can resist antibiotic onslaught, offering insight into strategies for combating drug-resistant pathogens.

The field of physiology is rich with examples of how membrane protein research has enhanced our understanding of health and disease. The implications extend to fields as diverse as nephrology, cardiology, and immunology. The ongoing investigations into the structural intricacies of membrane proteins continue to fuel our knowledge of these fundamental biological processes.

The Expanding Toolbox: Emerging Technologies in Membrane Protein Research

The excitement surrounding membrane protein research is sustained by an ever-evolving toolbox of cutting-edge technologies. These innovations are expanding the horizons of what we can uncover about these enigmatic molecules.

Nanodisc technology, for instance, has revolutionized the study of membrane proteins. Nanodiscs are lipid bilayer nanodiscs that allow membrane proteins to be solubilized and stabilized in a near-native environment. This technology has enabled researchers to overcome one of the major challenges in studying membrane proteins—keeping them stable and functional outside of the lipid bilayer.

Another breakthrough is the use of lipidic cubic phase for crystallizing membrane proteins. This method exploits the self-assembly properties of lipidic mesophases, allowing for the crystallization of otherwise challenging membrane proteins. It has opened doors to solving the structures of numerous membrane proteins, furthering our understanding of their functions.

Moreover, advances in labelling techniques have facilitated the investigation of dynamic processes involving membrane proteins. Techniques like solid-state NMR and electron paramagnetic resonance (EPR) spectroscopy have provided invaluable insights into conformational changes and interactions within the membrane.

These emerging technologies are akin to treasure maps, guiding researchers deeper into the labyrinth of membrane protein structures and functions. The ongoing refinement of these tools promises to uncover even more hidden gems in the years to come.

The Ethical Imperative: Responsible Conduct in Membrane Protein Research

While the quest to understand membrane proteins is a thrilling journey, it is also marked by ethical considerations. Responsible conduct in research is not only a moral obligation but a necessity to ensure the equitable distribution of benefits and to safeguard the interests of all stakeholders.

One ethical concern pertains to the use of animals in membrane protein research. Historically, animals have been employed for the production of membrane proteins and for studying their functions. However, this practice has faced ethical scrutiny, leading to the development of alternative methods, including cell-free expression systems and advances in computational modelling. These alternatives minimize the use of animals, reducing ethical concerns and advancing the field in a more humane direction.

Data sharing and collaboration also constitute essential ethical components of membrane protein research. The open sharing of data, models, and resources not only accelerates scientific progress but also ensures transparency and fosters a sense of collective responsibility within the research community. International collaborations, such as the worldwide Protein Data Bank (wwPDB), exemplify the commitment to ethical conduct within the field.

The ever-growing landscape of membrane protein research obligates us to confront ethical dilemmas and align our endeavors with the broader principles of fairness and responsibility.

Securing the Future: Funding and Resources for Membrane Protein Research

The ambitious pursuit of membrane protein research relies heavily on financial support and access to state-of-the-art resources. Funding agencies and research infrastructure play pivotal roles in propelling the field forward.

Funding opportunities abound, from government agencies to private foundations. The National Institutes of Health (NIH), the

European Research Council (ERC), and the Wellcome Trust are among the many organizations investing in membrane protein research. Researchers are encouraged to explore these avenues, develop competitive grant proposals, and secure the necessary resources to fuel their investigations.

In parallel, core facilities and research infrastructure are the backbone of membrane protein research. The availability of synchrotron facilities, cryo-EM equipment, NMR spectrometers, and computational resources is essential. Collaborative networks, like the Structural Genomics Consortium (SGC) and the Membrane Protein Structural Dynamics Consortium (MPSDC), offer access to cutting-edge technologies and promote collaborative endeavors. Researchers should leverage these resources to advance their work and ensure the continued growth of the field.

The Grand Needlepoint of Membrane Protein Research: A Tale Unfinished

The journey through the captivating world of membrane protein research continues to unfold. The impacts are tangible, shaping the future of drug development, our comprehension of health and disease, and the boundaries of scientific exploration. These dynamic molecules, intricately embedded in the membrane, are agents of transformation, charting the course of scientific understanding and medical innovation.

As we navigate the uncharted waters, armed with emerging technologies and a commitment to responsible research, the grand needlepoint of membrane protein research remains a tale unfinished. This journey, marked by ethical considerations, access to resources, and the unceasing quest for knowledge, promises to reshape the biomedical landscape for generations to come. It is a testament to the power of human curiosity, resilience, and collaboration, and it stands as a beacon lighting the path toward new discoveries and innovations in structural biology.

20.3 Encouragement for Future Researchers

In the hunt of understanding the complex world of membrane proteins, we find ourselves at the brink of unprecedented discoveries and advancements. The path forward is illuminated by the remarkable progress made in recent years, the innovative technologies that continue to emerge, and the increasing recognition of the crucial role membrane proteins play in diverse biological processes. As we embark on this journey, we must remember that while challenges persist, so do opportunities. This chapter serves as an encouragement to the future researchers who will undoubtedly shape the landscape of membrane protein research in the years to come.

Diversity Breeds Discovery

The vast diversity of membrane proteins, each with its unique structure and function, provides an endless source of potential discoveries. This diversity spans across various categories, including ion channels, transporters, receptors, and adhesion proteins. Each of these classes offers a distinct set of challenges and opportunities.

For instance, G-Protein Coupled Receptors (GPCRs), a subset of membrane proteins, have captured the attention of researchers due to their significance in signal transduction. The structural elucidation of GPCRs was considered an insurmountable task, but recent successes have paved the way for the exploration of their structure-function relationships. This is particularly encouraging for future researchers who will have the privilege of unearthing new insights into GPCRs and their roles in various physiological and pathological processes.

The Power of Cryo-Electron Microscopy (Cryo-EM)

Cryo-EM has emerged as a game-changer in the field of membrane protein structural biology. The technique's ability to capture high-resolution images of membrane proteins in their native environment is revolutionizing our understanding of their structure and function. Researchers can now investigate larger and more complex membrane protein assemblies, including multi-subunit complexes, which were previously beyond the reach of traditional methods.

Take the example of the bacterial flagellar motor, a remarkable molecular machine composed of numerous membrane proteins. Cryo-EM has enabled scientists to visualize the intricate details of this nanoscale apparatus, shedding light on its mechanism of action. Future researchers can harness the power of Cryo EM to explore equally complex membrane protein systems and contribute to our understanding of fundamental biological processes.

Synthetic Biology and Membrane Protein Engineering

As the field advances, the potential for membrane protein engineering becomes increasingly evident. Synthetic biology approaches offer the means to design and construct tailor-made membrane proteins with specific functionalities. For instance, the development of synthetic ion channels can have profound implications in drug delivery and disease treatment. This avenue of research encourages future scientists to explore the boundaries of what is possible in membrane protein design.

Moreover, membrane protein engineering opens doors to entirely novel applications. By reengineering membrane proteins for specific tasks, researchers can harness their intrinsic properties in biotechnology and nanotechnology. The intersection of synthetic biology and membrane protein research presents an exciting frontier that beckons innovators and pioneers.

Bridging the Gap between Structure and Function

While the elucidation of membrane protein structures has made significant strides, understanding their functions in the cellular context remains a challenge. The future of membrane protein research lies in the convergence of structural biology, biophysics, and cell biology. This interdisciplinary approach will allow us to decipher the intricate relationships between structure and function.

Researchers can draw inspiration from the study of potassium channels. The structure of these membrane proteins has been well-documented, but their functional dynamics are equally fascinating. By combining X-ray crystallography, electrophysiology, and live-cell imaging, scientists have begun to unravel the complex mechanisms that govern ion permeation and selectivity in these channels. Future researchers should continue to embrace such interdisciplinary collaborations to uncover the secrets hidden within other membrane proteins.

Drug Discovery and Targeted Therapies

The promise of membrane proteins in drug discovery is undeniable. Many currently approved drugs target membrane proteins, such as receptors and transporters. Yet, a vast landscape of untapped potential awaits exploration. The rising interest in allosteric modulators, which can fine-tune the activity of membrane proteins, underscores the richness of opportunities in this realm.

Consider the case of cystic fibrosis, a genetic disorder caused by mutations in the CFTR chloride channel. Researchers have made remarkable progress in developing drugs that correct the channel's function, offering hope to individuals with this debilitating condition. Such success stories underscore the potential for future researchers to discover novel drug targets and therapies within the vast tapestry of membrane proteins.

Ethical Considerations and Responsible Research

In the pursuit of knowledge, researchers must also embrace ethical considerations. The responsible conduct of research, particularly when working with membrane proteins, is paramount. Ethical research practices encompass animal welfare, data sharing, transparency, and adherence to regulations and guidelines.

As future researchers, it is imperative to prioritize ethical standards in all aspects of your work. Strive to minimize harm to research subjects, whether they are animals or human participants, and engage in rigorous and honest reporting of your findings. By doing so, you will not only contribute to scientific progress but also ensure the sustainability of the ever-evolving field of membrane protein research.

Collaboration and Knowledge Sharing

In an era marked by rapid information dissemination, collaboration and knowledge sharing have never been more critical. Future researchers should actively seek collaboration with peers, both within and outside their specific subfields. The exchange of ideas, techniques, and resources can lead to breakthroughs that might have remained elusive in isolation.

Consider the progress made in the study of membrane protein-lipid interactions. Researchers from various disciplines, including biophysics, lipidomics, and structural biology, have come together to unravel the complex interplay between lipids and membrane proteins. Future researchers should embrace this spirit of collaboration, which has the potential to accelerate discoveries in this multifaceted field.

Encountering the Unknown

The future of membrane protein research holds promise, but it is also characterized by the presence of the unknown. Every structure solved, every discovery made, unveils new questions and challenges. The pursuit of the unknown is an essential element of scientific research, and it is where future researchers will find both excitement and opportunity.

The field of membrane protein research is teeming with potential. As future researchers, you will contribute to the continued growth of this dynamic field, revealing the hidden intricacies of these vital biomolecules. Embrace the diversity, the power of new technologies, and the promise of engineering. Always remember the ethical responsibilities and the significance of collaboration. The journey ahead is not only about uncovering the mysteries of membrane proteins but also about expanding the frontiers of human knowledge and improving the quality of life through targeted therapies and drug discovery. With unwavering curiosity and determination, you have the opportunity to shape the future of membrane protein research and leave an indelible mark on the scientific tapestry of our time.

20.4 Final Thoughts on the Field's Potential

As we bring our exploration of the structural biology of membrane proteins to a close, it is fitting to ponder the wide-ranging implications and the limitless potential that this field holds for the scientific community, biomedical applications, and our fundamental understanding of life itself.

Illuminating the Biological Mysteries

Structural biology has been our beacon of light, illuminating the mysteries concealed within the complex lipid bilayers that guard the vital machineries of life. With growing momentum and burgeoning innovation, membrane protein research has not only broadened our horizons but continues to stretch them. In examining the very fabric of life's machinery at the atomic level, we have unravelled the intricacies of these proteins in their natural habitat – the cell membrane. This feat has empowered us to envision a future where fundamental biological processes and their deviations are understood with unprecedented clarity.

Take, for instance, the transport proteins that govern the flow of ions and molecules across membranes. The detailed structural insights into channels like the potassium channel Kv1.2 or the serotonin transporter have afforded us a profound understanding of their mechanisms. Such knowledge has not only unlocked therapeutic potential, but it also stands as a testament to the power of the membrane protein structural biology field in demystifying biological phenomena. The prospects for designing selective modulators and therapeutics targeting these proteins are brighter than ever before.

Biomedicine and Drug Development

The medical landscape, too, has reaped the benefits of our ceaseless efforts. G-protein-coupled receptors (GPCRs) serve as a prime example. With nearly 34% of drugs targeting GPCRs, the field has made inroads in both drug discovery and optimization. The crystal structure of the β2-adrenergic receptor – a GPCR – marked a turning point, offering us a glimpse into the architecture of these pharmacologically crucial proteins. This insight paved the way for the development of novel, more effective drugs for conditions ranging from cardiovascular diseases to psychiatric disorders.

Lipidomics has also witnessed a surge in popularity as we grasp the pivotal role of lipids in membrane protein structure and function. The research potential here is vast, encompassing lipid-protein interactions and lipidomic profiling for various biological contexts. We stand at the threshold of groundbreaking discoveries in lipidomics, including the identification of lipid biomarkers for diseases and the understanding of lipid-based mechanisms in cellular signalling.

Unravelling Cellular Signalling

Understanding cellular signalling is another promise on the horizon. When we peer into the molecular dialogues that govern vital processes, we uncover a wealth of insights that can redefine our approach to diseases. The structural elucidation of receptors such as the epidermal growth factor receptor (EGFR) and tyrosine kinases has elucidated how cell signalling is regulated at the molecular level. These revelations may lead to the development of more precise cancer therapies and open up new avenues for personalized medicine.

Emerging Technologies

Our journey wouldn't be complete without acknowledging the game-changing impact of emerging technologies. The advent of nanodisc technology, for instance, has transformed the way we investigate membrane proteins. By providing a stable and native-like environment for these proteins, nanodiscs have enabled more accurate structural studies. This innovation exemplifies how advancements on the periphery can fundamentally reshape the core of the field.

Lipidic cubic phase, often used for crystallization, has become a staple in our toolkit. Its ability to mimic the native lipid bilayer environment is invaluable. The cubic phase has granted us structural insights into some of the most challenging membrane proteins, pushing the boundaries of what we thought possible in structural biology.

We're also witnessing progress in the labelling techniques. Isotope labelling, site-specific labelling, and advances in NMR methodologies are offering finer insights into the dynamics and interactions of membrane proteins. Such techniques are essential as they enable us to capture snapshots of these proteins in their active states, shedding light on their functional mechanisms.

Crossroads of Disciplines

The growth of structural biology of membrane proteins has been greatly facilitated by the interplay between disciplines. Bioinformatics and computational modelling have provided indispensable tools for predicting structures, understanding interactions, and characterizing the dynamics of membrane proteins. As these fields continue to evolve, they will further augment the potential of membrane protein research.

The Ethical Imperative

Our journey would be incomplete if we didn't acknowledge the ethical dimension of our work. Ethical considerations are the bedrock of scientific integrity. They remind us that our pursuit is not just about deciphering the codes of life but also about the responsible stewardship of knowledge. As we forge ahead, our ethical compass should guide us in ensuring that the benefits of our discoveries are accessible, equitable, and used for the greater good.

The Path Forward

With infinite potential comes an infinite responsibility. As we contemplate the future of membrane protein research, we must remain vigilant in our commitment to rigorous science, ethical conduct, and collaborative efforts. Our understanding of membrane proteins is far from complete. We are in a continuous state of discovery, and each revelation uncovers more questions than answers.

The future holds an array of fascinating possibilities. Our structural insights could facilitate the engineering of designer membrane proteins tailored for specific functions. This prospect extends beyond therapeutic applications into biotechnology, where membrane proteins may serve as pivotal components in biosensors and other innovative devices.

The integration of membrane protein research with systems biology is another avenue. As we delve into the intricacies of cellular networks, understanding the role of membrane proteins in the broader context of cell physiology will be indispensable. This cross-disciplinary approach will provide holistic insights into how membrane proteins orchestrate cellular processes.

We should also keep a keen eye on recent breakthroughs in membrane protein biotechnology. From the development of high-throughput platforms for structural studies to the potential application of artificial intelligence and machine learning in data analysis, the field is on the cusp of transformative advancements. These technologies will not only accelerate research but also enhance our ability to predict and engineer membrane proteins.

In closing, our journey through the structural biology of membrane proteins has illuminated the molecular landscapes of life's barriers and gatekeepers. The promise of this field lies in its capacity to transform our understanding of fundamental biological processes, reshape biomedicine, and inspire ethical, collaborative, and innovative exploration. It is a journey that challenges us, empowers us, and beckons us to remain ever-vigilant in the face of an ever-evolving scientific landscape. As we continue our pursuit of knowledge, the potential for groundbreaking discoveries and transformative applications remains boundless. The story of membrane proteins is far from concluded; it is a narrative that continues to be written by researchers driven by a profound curiosity and a shared vision of a better, healthier future. The unfolding chapters are a testament to the enduring potential of this remarkable field.